AF307223

Zoophysiology Volume 36

Editors:
S.D. Bradshaw   W. Burggren
H.C. Heller   S. Ishii   H. Langer
G. Neuweiler   D.J. Randall

Springer-Verlag Berlin Heidelberg GmbH

Zoophysiology

L. Aitkin

Hearing – the Brain and Auditory Communication in Marsupials

With 43 Figures

Springer

Dr. Lindsay Aitkin
Monash University
Dept. of Physiology
Clayton, Melbourne
Victoria 3168
Australia

Cover illustration: "Outline drawing of the head of a sugar glider, from Fig. 5.1B."

ISBN 978-3-642-63705-6
ISSN 0720-1842

Library of Congress Cataloging-in-Publication Data

Aitkin, Lindsay.
   Hearing, the brain, and auditory communication in marsupials / L. Aitkin.
      p. cm. – (Zoophysiology; v. 38)
    Includes bibliographical references (p.   ) and index.
    ISBN 978-3-642-63705-6   ISBN 978-3-642-58739-9 (eBook)
    DOI 10.1007/978-3-642-58739-9
    1. Marsupialia – Physiology. 2. Hearing. 3. Auditory pathways.
   I. Title. II. Series.
   QL737.M3A58. 1998
   573.8'9192 — dc21

The use of general descriptive names, registered names, trademarks, etc. in this publication does not imply, even in the absence of a specific statement, that such names are exempt from the relevant protective laws and regulations and therefore free for general use.

Cover design: Design & Production GmbH, Heidelberg
Typesetting: Best-set Typesetter Ltd., Hong Kong

SPIN: 10542648   31/3137 – 5 4 3 2 1 0 – Printed on acid-free paper

For Joan

Preface

This monograph evolved from years of research into the auditory pathway and hearing of many species of marsupials. Its function is to give biologists, in particular neurobiologists, a broad description and review of what is known of the auditory sensory capacities and processing mechanisms in this large order of mammals. My initial interest in marsupials developed from collaborative work with Dr. Richard Gates at Monash and Melbourne Universities in the 1970s and was stimulated by curiosity as to whether concepts about the auditory system stemming from experiments mainly on domestic cats could be extended to mammals of other orders. My subsequent interest in Australian marsupials, aroused by collaboration with Dr. John Nelson at Monash University in the 1980s and 1990s, concerned their auditory systems and behavior *per se* and not as primitive cousins of eutherians. More recently, I have collaborated with Dr. Bruce Masterton at Florida State University in studies of New World marsupials. His sad death in 1996 has robbed neurobiologists of one of our most provocative thinkers and hypothesis testers.

I would like to thank the Department of Physiology at Monash University for making many facilities available to me, the National Health and Medical Research Council of Australia and the Australian Research Council for providing funds for research, and Jill Poynton and Michelle Mulholland, who illustrated this volume. Over the years I have greatly benefited from discussions about marsupials and their idiosyncrasies with Karen Glendenning, Peter Heil, Jack Pettigrew, and Ramesh Rajan. In particular, I wish to thank Mike Calford and John Nelson for their comments on this manuscript.

Melbourne 1997 *Lindsay Aitkin*

Contents

Chapter 1

Introduction

1.1 Why Study Marsupials?

Marsupials belong to one of the three orders of class Mammalia: Eutheria, Marsupialia, and Monotremata. Marsupials are distinguished primarily by their method of reproduction, in which the embryo is born at a very early stage and continues its development external to the uterus, in the absence of a chorioallantoic placenta. Thus, eutherian mammals are more commonly called placental mammals, although the early uterine phase is associated with a choriovitelline placenta. Short (1985) writes: "To suggest that marsupials do not have a placenta belies the evidence of our eyes ... it is no longer justifiable to use the term 'placental mammals' as a synonym for the Eutheria". However, although the term "Eutheria" has its own misleading connotations (see below), it will be used in the present account.

Most female marsupials are easily distinguished from eutherians by the possession of an abdominal pouch, or marsupium, which is best developed in climbers, hoppers, diggers, and swimmers. In some species of the Didelphidae and Dasyuridae, the "pouch" is simply a few folds of skin on the abdomen around the nipples, affording the developing young some protection. A number of other characteristics differentiate marsupials from eutherians and monotremes – the differing arrangements and number of teeth in the upper and lower jaws, the different arrangements of bones in the skull, which in marsupials has a large facial area and small cranial area, and the lack of a corpus callosum connecting the two cerebral hemispheres. Interestingly (from an auditory neurobiologist's perspective), there is a distinctively marsupial arrangement of the cochlear nuclei in the medulla of the brain (see Chap. 5).

Very little research has been carried out on the brain and hearing of marsupials compared with the great volume concerning eutherians, a state which deserves to be rectified. Some research on marsupials has been prompted by the dubious concept of 'primitiveness' (see below), whereas other researchers have concentrated on the adaptive specializations of certain marsupial species. Still further studies have compared the behavioral and ecological convergence between marsupials and eutherians (e.g., Lee and Cockburn 1985). In particular, investigations into the different reproductive mechanisms and behavior of marsupials (e.g., Tyndale-Biscoe 1973) have cast light on mammalian reproduction in general. Embryonic and fetal development is, relative to eutherians, easy to study in the externally developing marsupial. Determinants of development can be manipulated and altered with little interference to the mother or other members of the same litter.

This books is concerned with hearing and communication in marsupials, a subject poorly studied but deserving some consideration since many marsupial species make important use of hearing for prey detection, predator avoidance and conspecific communication. A number of the arboreal species are highly vocal, with a rich repertoire of communication sounds. Knowledge of the hearing and vocalizations of marsupials may be very helpful in understanding the ecology of these mammals. Chapters in this work deal with the frequency hearing ranges determined by behavioral and physiological means; the vocal behaviors and their relationships to hearing ranges and habitats; aspects of the anatomy and physiology of the auditory system from ear to cerebral cortex; and the development of the auditory nervous system in comparison with that in the more widely studied eutherians.

1.2 Evolutionary Considerations

When and where did marsupials first appear and from what ancestors? The earliest marsupial fossils occur in the late Cretaceous of North America; those in South America occur mostly later (with a few exceptions in the Cretaceous), in the late Paleocene (Table 1.1). Those of Australia are even more modern. Until recently, the oldest Australian marsupial fossil (Wynyardia; Table 1.1) was identified in the early Miocene of Tasmania (Simpson 1945; Ride 1962; Kirsch 1977).

Table 1.1. The geological succession in relation to marsupials

Time	Era	Period	Event – biological and geological
10000 years B.P.	Quarternary	Holocene	Modern man established
2 Ma		Pleistocene	*Homo sapiens* appears
5 Ma	Tertiary	Pliocene	N. America recolonized by marsupials
23 Ma		Miocene	Carnivorous marsupials, S. America (extinct) Most recent Riversleigh marsupial deposits (Australia) Caenolestoidea (S. America) Wynyardia (Tasmania)
35 Ma		Oligocene	Earliest Riversleigh deposits "Polyprotodonts" elsewhere in Australia N. and S. American didelphoids European didelphoids
56 Ma		Eocene	Australia separates from Gondwanaland
65 Ma		Paleocene	Most S. American marsupials
146 Ma	Mesozoic	Cretaceous	Modern America and Australia connected as Gondwanaland Earliest marsupial fossils (N. America) First eutherians Monotremes (Australia)

However, major finds such as those of the early 1980s at Riversleigh in North Queensland have provided evidence that marsupials were present in the Oligocene period (Archer et al. 1991; Rich 1991). These fossils are very similar to modern forms and are quite different to those of North America, suggesting that there may be a much older phylogeny in Australia than presently known. Kirsch (1977) suggested that the lack of earlier mammal fossils in Australia may be due to the unsuitability of the geology of that country for the preservation of Tertiary fossils. However, fossils of other vertebrates extend back to the early Cretaceous (Archer et al. 1991). Ride (1968) argued that the lack of knowledge about the distribution of mammals in Australia (up to the late 1960s) was due to a lack of zoologists!

Marsupials, particularly North American opossums, are often considered primitive. Much of this belief is based on the life styles and appearance of opossums, although there is some evolutionary reasoning behind this view. Fernandez and Schmidt (1963) wrote: "opossums (Didelphidae) are the most primitive of the contemporary marsupials. These animals are 'living fossils' and have changed little since Cretaceous times" (p. 157). The concept of primitiveness can be traced back over 100 years to T.H. Huxley's (1880) introduction of the terms Prototheria (monotremes), Metatheria (marsupials), and Eutheria ("placentals"), suggesting a serial relationship in terms of evolution between these mammals (see Tyndale-Biscoe 1973, for review). However, examination of a number of features of the biology of modern mammals and of the anatomical features of fossil specimens suggests that monotremes, marsupials, and eutherians have all evolved independently from common, mammal-like reptiles (e.g. Ride 1962). In this sense marsupials are no more primitive than eutherians, although the extent of subsequent change from the ancestor is another factor to be considered.

Since the time of mammal-like reptiles, the skull expanded anterior to the ear as the brain grew. Different component bones of the skull in the three mammalian orders are responsible for the expanded brain case; the result is much the same, but the anatomical changes producing it are different in monotremes, marsupials, and eutherians. In relation to hearing, different bones have been employed to construct the floor of the epitympanic recess (middle-ear cavity) in marsupials and eutherians. Collectively, data such as these suggest that monotremes have evolved independently of marsupials and eutherians prior to the emergence of the latter orders and that eutherians could not have evolved from marsupials. Certainly, the fossil record reveals the concurrent existence of monotremes, marsupials, and eutherians during the later Cretaceous (Table 1.1).

Thus, for example, fossil specimens collected from Alberta, Canada, and studied with radiographs and scanning electron microscopy include the oldest known ear structures of any therian fossils (Meng and Fox 1995a) and include several marsupials. They demonstrate the existence during the Late Cretaceous of fully coiled cochlear canals and the development of the radial system of the cochlea. Material from both marsupials and placentals in Montana, also Late Cretaceous (about 65 million years B.P.), show considerable similarities between the two orders at that time (Meng and Fox 1995b; Fig. 1.1, from Meng and Fox 1995a). The inner ear structures of these therians differ from modern mammals in having a shorter basilar membrane and fewer turns in the cochlea.

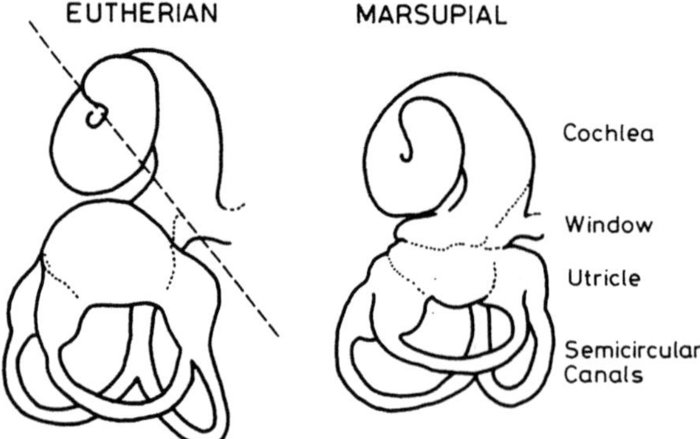

Fig. 1.1. Reconstructions of the osseous labyrinth from Late Cretaceous placental (*left*) and marsupial (*right*) fossils. The cochleas (*upper*) both have about 1.5 turns; the modiolar axis is indicated by the *dashed line, left*. (Drawing based on Fig. 3, Meng and Fox 1995a)

What is the zoogeography of marsupials and how did the current distribution evolve? The earliest fossils are found in North America; evidence suggests that marsupials may have spread into the Eurasian continent but became extinct there and in North America in the Miocene (Ride 1962). Toward the end of the Pliocene there may have been a reinvasion from South America when the land bridge was established. The earlier suggestion that the arrival of marsupials in Australia could have been via Asia ("island hops", Simpson 1965) is difficult to reconcile with evidence that land bridges connecting Australia and Asia have only been in existence for some 20 Ma. More credence is now given to the concept of southern land bridges connecting Australia and America as parts of Gondwanaland; populations of marsupials migrated from South America to Australia across the resultant land bridge (Archer 1984; Ride 1962). Following the ice ages, during the late Eocene, the marsupial populations of Australia were isolated from competition with Eutheria, so they continued in Australia to fill the ecological niches occupied in the Americas and elsewhere by eutherians. The marsupials of North America became restricted to one family, Didelphidae, and these animals, opossums, are considered by many to be particularly primitive marsupials compared with those of Australia. According to Kirsch (1977, p. 19):

... "the fossil history of marsupials is as long as, but probably no longer than, that of placentals; geographically, marsupials were probably restricted to South America and Australia after the Miocene and until the Panamanian land bridge was reestablished. They proliferated in parallel ways on those two continents, each radiation producing a prodigious variety of carnivorous and diprotodontid forms. In spite of the similarities, the two radiations differed significantly in that the Australian fauna was a complete group including herbivores, while in South America marsupials tended to complement the placental herbivores with carnivorous and rodentlike species. The fossil history of South America suggests that the marsupial families there differentiated early, but the Australian record is too poor to indicate if the radiation on that continent paralleled the South American temporally or postdated it. The great distinctiveness and diversity of the Australian diprotodonts suggest that the major groups were established early in Australia as well, however."

1.3 Taxonomic Considerations

The accepted scheme of classification, that of Linnaeus, provides each species with a specific name preceded by a generic name, shared by closely related species. Thus, the North American opossum is *Didelphis virginiana*; a species related in a host of details but found in other parts of America is *Didelphis marsupialis*. Genuses with common characteristics are grouped together in families – thus, *Didelphis* and the South American opossum genus *Monodelphis* both belong to the Didelphidae. Similar families belong to an order, e.g., Marsupialia. However, because of the ever-increasing numbers of character sets used in taxonomic analyses (e.g., Johnson et al. 1994) and the new discoveries being made about the fossil record (e.g., Archer et al. 1991), the classification of generic marsupial groupings into families, superfamilies, and orders is in a continuing state of flux. This book is not directly concerned with deciding upon a "correct" scheme, but it is of interest to discuss briefly the different schemes of marsupial classification.

Of the extant species of marsupials, about one third live in the Americas and two thirds in Australasia. In Walker's (1968) simple scheme, the order *Marsupialia* is composed of 8 families: Didelphidae (omnivores), Dasyuridae (carnivores), Peramelidae (insectivores), Notoryctidae (insectivore), Caenolestidae (insectivores/carnivores), Phalangeridae (arboreal herbivores), Phascolomidae (ursine herbivores) and Macropodidae (terrestrial herbivores). Kirsch and Calaby (1977) list a single order with sixteen families; other classifications are still more detailed (e.g., Woodburne 1984; Aplin and Archer 1987). A number of subdivisions have been suggested at various levels, including two suborders, polyprotodonts and diprotodonts (e.g., Johnson et al. 1994), named for an obvious morphological feature: dentition (Tyndale-Biscoe 1973). The Polyprotodonta have long snouts and possess many simple, sharp-pointed teeth, including four or five incisors in the upper jaw and three in the lower jaw. This suborder includes the families Didelphidae, Dasyuridae and Peramelidae. The remaining marsupials have a more complex dentition, with three incisors in the upper jaw and one large procumbent incisor in each dentary; these marsupials are the Diprotodonta. Foot structure is a useful classificatory indicator – *syndactyly*, where digits are fused functionally together, versus *didactyly*. Animals that show syndactyly may be either poly- or diprotodontid; in general, the more characteristics used for classification, the more valuable the distinction.

Simpson (1945) subdivided the order Marsupialia into 6 superfamilies and 13 families, and a recent review by Clemens et al. (1989) amplifies further the Simpson scheme. Most of the modern American opossums are collected into the family Didelphidae; the small South American families Caenolestoidae and Microbiotheriidae are distinctly different from the Australian marsupials and serologically more similar to the Didelphidae. On the other hand, the South American opossum *Dromiciops australis* is the nearest relative of the Australian marsupials (Fig. 1.2), being more similar to them than to the American opossums (Szalay 1982); this close relationship has prompted the inference that Australian marsupials evolved from a South American ancestor (e.g., Flannery 1994).

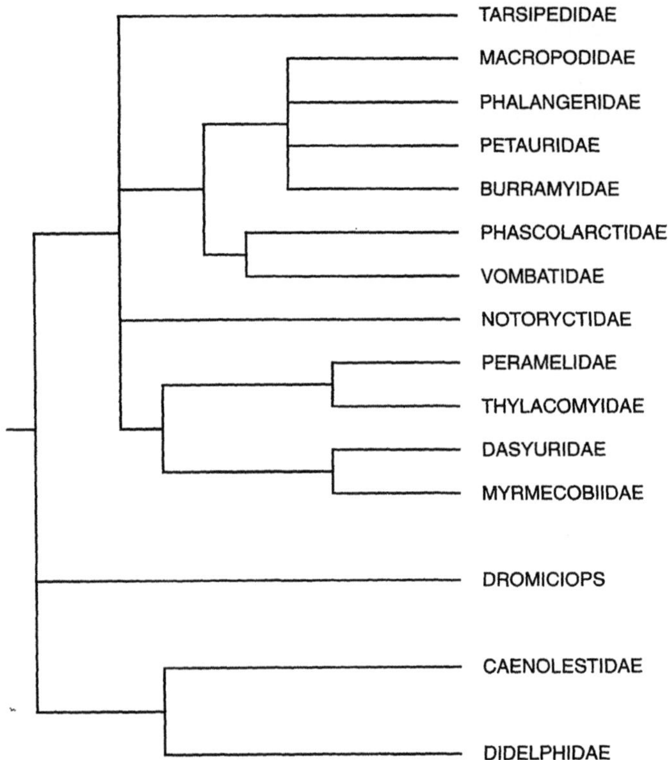

Fig. 1.2. The serological affinities of most of the extant marsupial families. (Modified from Kirsch 1977)

The marsupials of Australasia are divided into one minor and three major lineages, Dasyuroidea, Perameloidea, Diprotodonta and Notoryctidae (marsupial mole). The three superfamilies are further subdivided: **Dasyuroidea** into Dasyuridae (e.g., quolls), Myrmecobiidae (numbats), and Thylacinidae (marsupial "wolf", now extinct); **Perameloidea** into Peramelidae (bandicoots) and Thylacomyidae (bilbies); and **Diprotodonta** into Vombatidae (wombats), Phascolarctidae (koala), Macropodidae (kangaroos and wallabies), Potoroidae (potoroos)(the latter two families are usually considered together), Phalangeridae (brushtail possums, cuscus), Burramyidae (pigmy possum, feathertail glider), Petauridae (gliders), Pseudocheiridae (ringtail possums)(the latter two families are usually considered together), and Tarsipedidae (honey possums). Serological studies show affinities between the Australian groups, with **Diprotodonta** collected together (excepting the Tarsipedidae) and the Didelphidae collected together; both are separate from the Caenolestidae (Kirsch 1977).

The most precise taxonomic relationships among extant marsupials are determined by comparisons of DNA and chromosomal morphology and length, and by serological means, in which the similarities of serum proteins are compared using the antigen-antibody reaction. Figure 1.2 (modified from the review of Kirsch 1977) summarizes the relationships between the various marsupial families as

6

determined from serological studies. Note that only the horizontal axis has dimension, with linkages of groups successively less close from right to left.

1.4 Ecological Considerations

Marsupials can be classified in relation to their principal diets, i.e., as carnivores, insectivores, herbivores, and omnivores and in relation to their habitats, i.e., terrestrial, sub-terrestrial, arboreal, and aquatic species. Lifestyles can be equated easily between members of different orders of mammals. Even the shapes of the heads of many grazing herbivorous kangaroos and wallabies resemble those of deer; the various arboreal possums in Australia occupy niches filled by squirrels in Europe and North America; insectivorous marsupial "moles" and platypus share many of the features exhibited by European and African moles and insectivores; and the carnivorous Dasyurids combine attributes of mustelids, such as weasels and ferrets, and cats. The marsupials of Australia provide a full spectrum of ecological variants, surpassing the variety seen on the American continents with the exception of the aquatic *Chironectes minimus* of South America.

Those marsupials belonging to the **Dasyuroidae** are predominantly carnivorous in the sense that they kill and eat warm-blooded prey. These include the largest living marsupial carnivore, the Tasmanian Devil (*Sarcophilus harrisii*), the quolls ("native cats", a species of *Dasyurus*; Fig. 1.3A), various species of *Antechinus* (marsupial "mouse", Fig. 1.3B), the desert-living kowari (*Dasyuroides byrnei*; see Fig. 5.1D) and dunnart (*Sminthopsis* sp.; Fig. 1.3C). Some South American marsupials are also primarily carnivorous (e.g., the four-eyed opossum, *Philander* and the water opossum, *Chironectes*).

Many of the arboreal Didelphidae (such as the Virginia opossum, *D. virginiana*; Fig. 1.4A) and Phalangeridae (e.g., the brushtail possum, *Trichosurus vulpecula*; Fig. 1.4B) are highly adaptable in their diets. Although they primarily eat foliage, fruit, and seeds, they are at times insectivorous and even carnivorous. It is likely that the dietary adaptability of the Virginia opossum and brushtail possum accounts for their success in a variety of environments in America and Australia, respectively. Brushtail possums were introduced into New Zealand many years ago and have now become a major pest in that country. Other arboreal Australian species are mainly vegetarian. Some, such as the feathertail glider (*Acrobates pygmaeus*; Fig. 1.4C), also eat insects, whereas others, such as the common ringtail possum (*Pseudocheirus peregrinus*; Fig. 1.4D) and the koala, are almost exclusively folivorous.

Terrestrial herbivores include the macropodids (e.g., kangaroo; Fig. 1.5A), potoroos, and the wombat (Fig. 1.5B). Other terrestrial forms are omnivorous; many eat insects, such as the bandicoots (Fig. 1.5C), caenolestids, the opossums *Monodelphis domestica* (see Fig. 5.1E) and *D. australis*, and the marsupial mole, *Notoryctes typhlops*. The insectivorous striped possum (*Dactylopsila trivirgata*; Fig. 1.5D) has a specialized digit (arrow, Fig. 1.5D) which it uses to search in cavities in wood for insects; in this respect it has a remarkably similar behavior to the Aye Aye lemur of Madagascar.

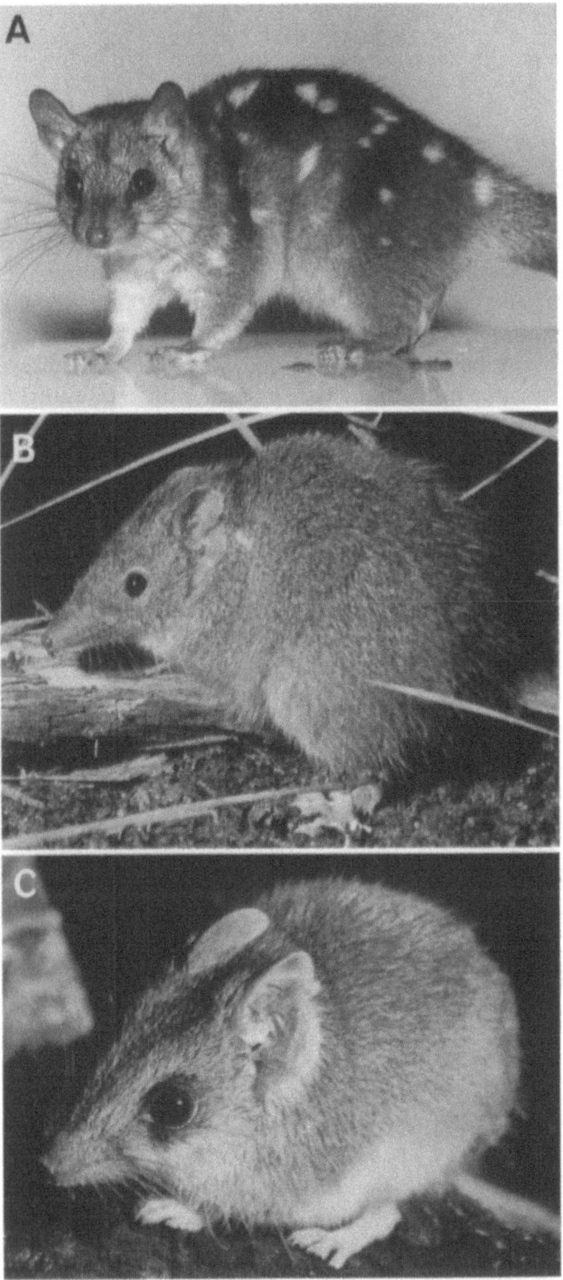

Fig. 1.3A–C. Three members of the Dasyuridae. **A** Northern quoll (*Dasyurus hallucatus*), head–body length 200 mm. **B** Antechinus (*Antechinus stuartii* or *swainsonii*), head–body length 100 mm. **C** Dunnart (*Sminthopsis leucopus* or *murina*), head–body length 90 mm. Acknowledgment is made to Dr. J. Nelson (**A**) and the Mammal Survey Group of Victoria (**BC**)

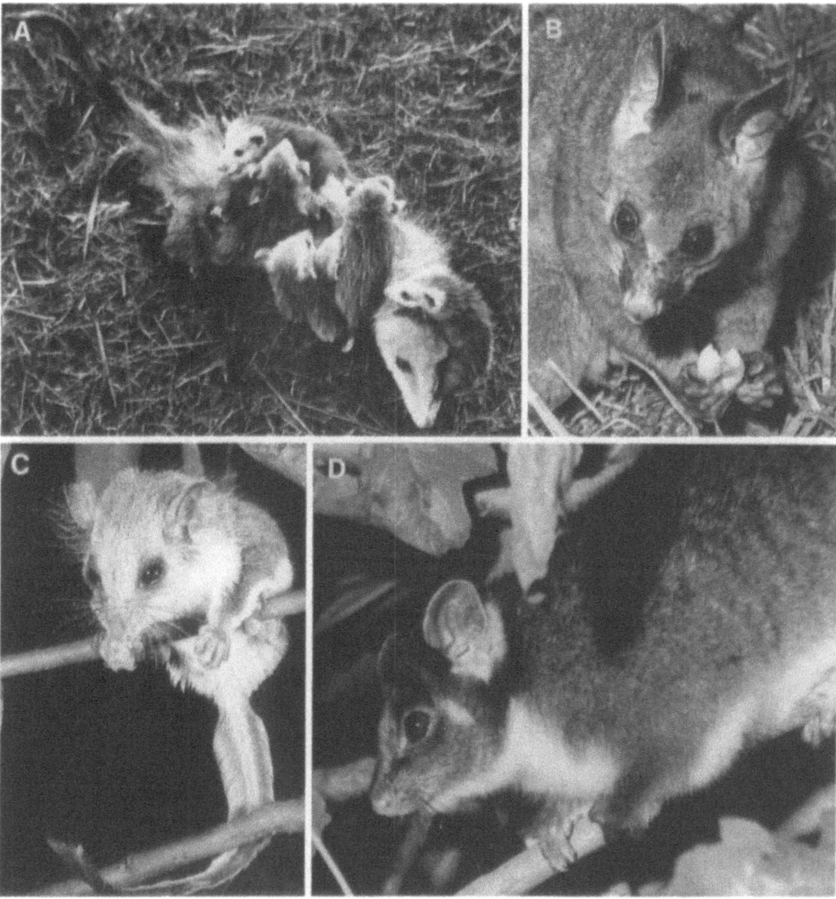

Fig. 1.4. **A** Virginia opossum and litter of young (*Didelphis virginiana*), head–body length 500 mm. **B** Brushtail possum (*Trichosurus vulpecula*), head–body length 500 mm. **C** Feathertail glider (*Acrobates pygmaeus*), head–body length 66 mm. **D** Ringtail possum (*Pseudocheirus peregrinus*), head–body length 350 mm. Acknowledgment is made to the late Dr. R.B. Masterton (**A**) and the Mammal Survey Group of Victoria (**D**)

Most Australian marsupials are nocturnal or crepuscular; very few are diurnal. The hot, dry summer days in many parts of Australia would represent a challenge for biological water conservation; smaller mammals would also be easy targets for diurnal raptors during the day. The tendency for animals to be active during the night imposes special requirements on the visual system and favors the use of hearing as the most effective distance sense.

The successful adaptation of a species to its environment requires, among a host of other factors, an auditory system sensitive to the airborne vibrations generated by its conspecifics, adult and juvenile, by its prey and predators, and by abiotic factors such as the sounds made by wind and water. This ensemble of acoustic features has been termed the *acoustic biotope* (Aertsen et al. 1979). The

9

Fig. 1.5. **A** Eastern grey kangaroo (*Macropus giganteus*), head–body length 1500 mm. **B** Hairy-nosed wombat (*Lasiorhinus latifrons*), head–body length 900 mm. **C** Short-nosed bandicoot (*Isoodon obesulus*), head–body length 350 mm. **D** Striped possum (*Dactylopsila trivirgata*), head–body length ca. 250 mm. Acknowledgment is made to the Mammal Survey Group of Victoria (C)

factors influencing the biological elements in this ensemble will relate to a species' diet (e.g., carnivore vs. herbivore), size (small vocal tracts generally relate to higher pitched vocalizations), and habitat (terrestrial, arboreal, aquatic, etc.). The auditory systems of species of varied phylogenetic backgrounds but with similar diets, sizes, and habitats are likely to have requirements that bear a greater similarity than those of identical phylogenetic backgrounds but diverse lifestyles. There is thus no reason to expect that marsupials will have auditory systems whose functions differ dramatically from those of eutherians. However, one cannot predict a priori that functionally similar systems will have similar structures.

1.5 The Approach of this Volume

In Chapter 2 a brief description and analysis are made of the structure and physiology of the cochlea and the brain auditory pathways of mammals, derived from work carried out mainly on eutherian species. In Chapter 3 the hearing ranges of marsupials, which have been determined by behavioral means or with electrophysiological methods, are examined. An analysis of the vocal behaviors follows in Chapter 4. Attention here is given particularly to the relation between hearing range and the spectra of calls to which each species must respond.

Chapter 5 discusses the designs of the peripheral auditory systems of marsupials in comparison with eutherians. It will be shown that, even at this level of the auditory system, differences exist between the orders. The organization of the brain auditory system is the subject of Chapters 6 and 7. There are distinctively "marsupial" traits in the arrangement and wiring of auditory nuclei, from cochlear nuclei to the cerebral auditory areas. The dearth of work on the physiology of neurons in the marsupial auditory system makes it difficult to identify correlative functional differences from eutherians, but some suggestive material is available from studies of the auditory midbrain and cortex.

Finally, in Chapter 8, the development of hearing in marsupials is described; discussion allows major differences to eutherian auditory systems to be easily distinguished. Furthermore, the accessibility of the developing marsupial makes the application of physiological and anatomical techniques in studying the development of hearing very early in life, equivalent to the fetal phase of eutherians, relatively easy. Many of the determinants of marsupial development will also apply to all mammals.

Chapter 2

The Design of the Mammalian Auditory System: A Brief Overview

There are three stages in sound reception at the periphery. First, sound reaches the external ear, or pinna, where, owing to the difference in acoustic impedance between air and tissue, most of the sound energy is dissipated by reflection. The sound entering the ear canal will be amplified to an extent dependent on the direction of incident sound waves in relation to the acoustical axis of the pinna. Thus, the pinna acts as a directional (and, for some species, highly mobile) sound amplifier.

The length and diameter of the ear canal will provide a frequency-selective resonant amplification, and the eardrum will vibrate, imparting motion to the auditory ossicles (Fig. 2.1). Due to the fact that the eardrum is usually much larger in area than the footplate of the third middle ear bone, the stapes, the pressure received by the eardrum is amplified before it enters the cochlea. This amplification is enhanced by a lever action of the ossicles. In essence the middle ear acts as an impedance-matching sound amplifier, solving the problem of the transfer of airborne sound vibration to a fluid-based transmission.

In all mammalian ears, the basic receptor elements, hair cells, are supported on a moveable platform, the basilar membrane (Fig. 2.2). This whole assembly is immersed in a fluid-filled vesicle, the cochlea, a complex spiral cavity within the temporal bone. The hair cells are stimulated by vibrations passing through the membrane systems, the basilar membrane and Reissners membrane. These membranes separate three fluid compartments along the length of the cochlea (scalae vestibuli, tympani, media).

The stereocilia (hairs) of the hair cells are bent due to the differential motion of basilar membrane and fluid. This displacement opens channels near the tips of stereocilia, allowing mainly potassium ions to enter the cell. The ensuing depolarization of the hair cell causes the release of a chemical transmitter which depolarizes the attached cochlear nerve afferents.

2.1 Structure and Function of the Organ of Corti

The organ of Corti consists of the basilar membrane, upon which sit the three rows of outer (OHC) and one row of inner (IHC) hair cells; a variety of supporting cells including border cells, inner phalangeal and inner pillar cells for the IHC and outer pillar cells; Deiters and Hensens cells for the OHC; and the unmyelinated endings of cochlear nerve fibers (Fig. 2.2; see, e.g., Friedman and Ballantyne 1984). The cochlear nerve contains afferent and efferent fibers, the latter being

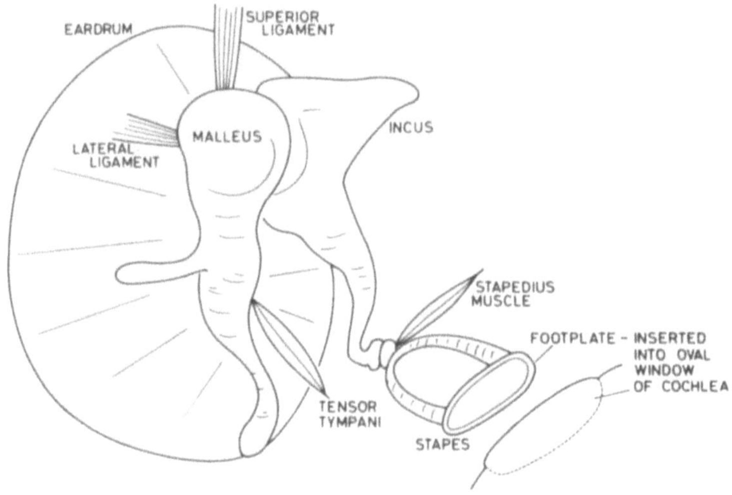

Fig. 2.1. Middle ear structures in a generalized eutherian mammal, viewed from within the middle ear

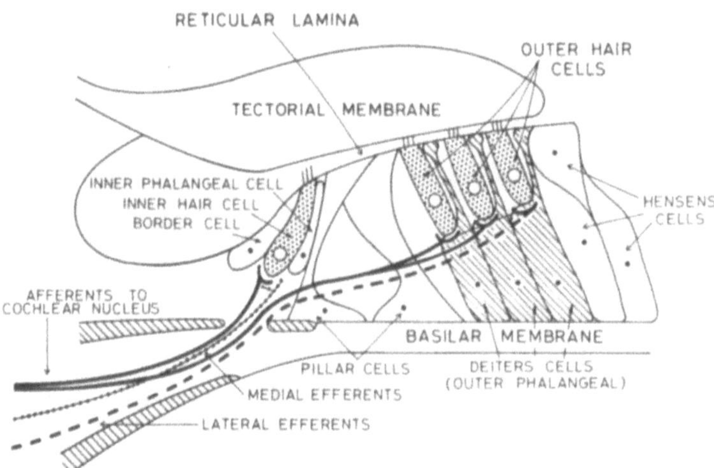

Fig. 2.2. Elements of the organ of Corti of a mammal as seen in a section at right angles to the cochlear spiral

greatly outnumbered by the former. Most of the afferents synapse with the IHC (95% in cats; Spoendlin 1984), with the remainder supplying the much more numerous OHCs. This indicates that the IHCs are the primary transducers of sound waves. Efferent fibers synapse mostly on the OHC directly (lateral efferents); a smaller number synapse on IHC afferent dendrites (medial efferents). Another difference between OHCs and IHCs lies in the contractile properties of OHCs, absent in IHCs (Brownell et al. 1985). The direct efferent innervation of electromotile OHCs allows the possibility that mechanical events at the cochlea can be actively regulated by the OHCs.

14

There are differences in the efferent innervation of the base (the end of the cochlea nearest the stapes) and the apex; the direct efferent synapses on basal OHCs are rare or absent at the apex. This resembles an early stage of development at the base (see, e.g., Pujol et al. 1997) and the apical region is often considered "primitive". The lack of efferent innervation to apical OHCs suggests that this part of the cochlea responds only passively to sound, unlike the "active" base.

Studies using a variety of techniques have shown that a traveling wave passes along the basilar membrane, reaching peak amplitude at a place related systematically to the frequency of the stimulating sound wave (for review, see Dallos 1984). The classic experiments were carried out by von Békésy (1960) on scale models, human cadavers, and a variety of live eutherian ears. The basic frequency or tonotopic organization of the cochlea is conferred by the continuous change in the mechanical properties of the basilar membrane from the base, where high frequencies cause maximum displacements, to the apex, where low sound frequencies are effective. This tonotopic organization is refined by active involvement of the electromotile OHCs. Damage to or poisoning of these cells greatly attenuates cochlea frequency selectivity. The net neural result is that cochlear nerve fibers are sharply tuned to different frequencies and are tonotopically organized within the nerve as they leave the cochlea.

2.2 Cochlear Potentials

It has long been possible to record from the round window of the cochlea and measure the amplitude, frequency selectivity, and threshold of potentials generated by cochlear processes. In some species, such as cats and guinea pigs, the technique is relatively simple: a hole is made in the auditory bulla after it has been exposed by surgical removal of the overlying scalp and temporal muscles. The round window can usually be seen through the hole (depending on the species), and a low-impedance wire electrode is allowed to rest on the round-window membrane. Acoustic stimuli evoke two major potentials: the cochlear microphonic (CM, the sum of receptor potentials generated by hair cells) and the compound auditory nerve action potential (CAP or N1), the sum of cochlear nerve action potentials evoked by the stimuli (see, e.g., Dallos 1984). The electric circuitry is in fact quite complex, and it is not a simple matter to identify the precise location along the cochlea of the elements giving rise to the potentials. The result of this is that the amplitude of the CM is not simply related to the number of hair cells activated, particularly at high frequencies. Consequently, the "CM audiogram", determined by measuring the sound-pressure levels needed to evoke a certain CM potential amplitude as a function of frequency, has only limited value, mainly for the lower frequency range of hearing. Furthermore, because CM potentials are graded continuously as a function of sound-pressure level, the CM has no real "threshold".

2.3 Anatomy of Auditory Nerve and Brainstem

Fibers from the cochlear spirals assemble in an orderly fashion, so preserving spatially the high-to-low frequency (tonotopic) organization of the basilar membrane. These fibers proceed through the internal auditory meatus and, upon reaching the brain, bifurcate into anterior and posterior branches. These supply the anteroventral (AVCN) cochlear nucleus, and the posteroventral (PVCN) and dorsal (DCN) cochlear nucleus, respectively (Fig. 2.3; Lorente de No 1933; Willard 1993). Fibers from the cochlear nucleus proceed to the medial and lateral superior olives (MSO, LSO) and medial nucleus of the trapezoid body (MTB) and also, along with fibers from the superior olive, to the midbrain nuclei – the inferior colliculus and the nuclei of the lateral lemniscus (NLL; Fig. 2.3). Axons from the midbrain project to the medial geniculate nucleus and the auditory cortex in the temporal lobe of the neocortex.

The superior olive is the source of the olivocochlear bundle, a pathway composed of efferent fibers destined for the inner and outer hair cells of the cochlea. These efferent pathways have been well described in eutherian laboratory animals (for review see Rajan 1990; Warr 1992) and can be divided into a lateral system supplying the outer hair cells, originating from cells in and around the LSO, mostly on the ipsilateral side and a medial system, originating from cells in the medial part of the superior olive, mainly on the contralateral side.

The term "auditory midbrain" describes the assemblage of nuclei in the midbrain receiving auditory input (Fig. 2.3) and includes the subdivisions of the IC, the deep layers of the superior colliculus, the dorsal nucleus of the lateral lemniscus, and smaller structures, such as nucleus sagulum (for review, see e.g. Aitkin 1986). It is dominated by the *central nucleus of the inferior colliculus (CNIC)*.

Anatomical and physiological work carried out on the inferior colliculus in eutherians has shown that CNIC has a pivotal role in the channeling of information and its rerouting to different thalamocortical and midbrain pathways. In brief, CNIC is the origin of the primary auditory pathway proceeding through the medial geniculate body (MG) to the auditory cortex, important for auditory perception; the external nucleus of the inferior colliculus (EN) receives additional input from the somatosensory system and is the site of origin of a number of acousticomotor pathways; and the dorsal cortex of the inferior colliculus (DC) is likely involved in auditory attention. The dorsal nucleus of the lateral lemniscus (NLL in Fig. 2.3) and the deep layers of the superior colliculus are important for the localization of sound, in the case of the latter structure for the correlation of auditory and visual spatial information.

A multiplicity of pathways originates from the IC of eutherians (for review, see Aitkin 1986). The best understood and probably largest of these is the lemniscal pathway, involved in auditory perception, reaching the auditory cortical areas through the medial geniculate body. This is comprised of component pathways originating from the different divisions of the IC (mostly CNIC), both ipsilateral and (to a lesser extent) contralateral. A second group of connections serve motor responses to sound. Some motor connections (e.g., to middle ear muscles) are made at the level of the brain stem, whereas others (via the superior colliculus, reticular formation, or pontocerebellar systems) effect orientation reflexes of the

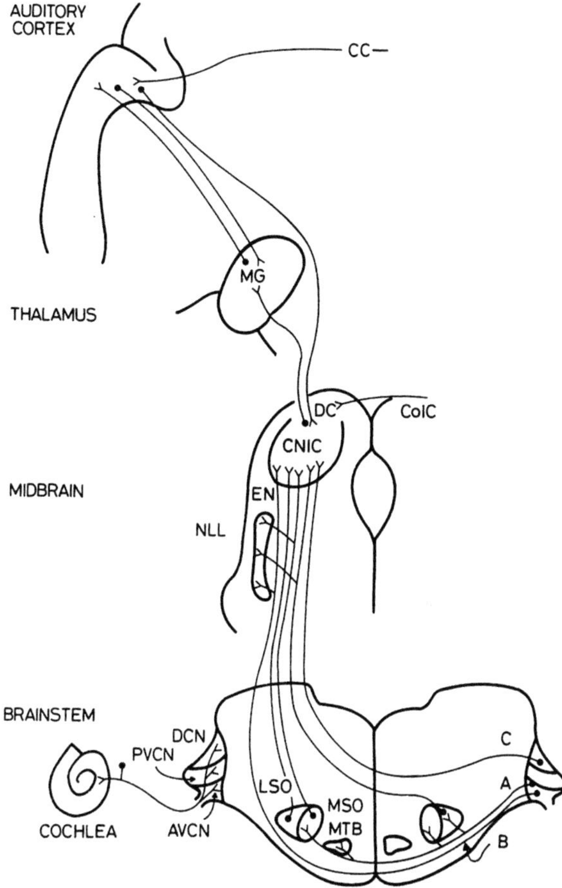

Fig. 2.3. Connections of the auditory pathway in a mammal, viewed in slices taken at the levels of brain stem, midbrain, thalamus, and cerebral cortex. Cochlear nerve fibers branch on entering the cochlear nucleus and supply the anteroventral (*AVCN*), posteroventral (*PVCN*), and dorsal (*DCN*) cochlear nuclei. *Pathway A*, issuing from *AVCN*, supplies terminals to the lateral (*LSO*) and medial (*MSO*) superior olivary nuclei of the same side and the MSO and medial nucleus of the trapezoid body (*MTB*) of the opposite side. Cells in the LSO and MSO project axons to the central nucleus of the inferior colliculus (*CNIC*); those from the MSO are largely ipsilateral (as shown), whereas those from the LSO are largely contralateral (not shown). Collateral branches supply the nuclei of the lateral lemniscus (*NLL*). The *CNIC* is the focus of ascending projections – the above-mentioned pathways plus pathways *B* (direct from *AVCN*), *C* (from *DCN*), and afferents from *NLL* terminate here.

From the inferior colliculus, including its external nucleus (*EN*) and dorsal cortex (*DC*), fibers ascend to the medial geniculate body (*MG*) and ultimately the auditory cortex. Pathways descend from the cortex to the thalamus and colliculus; cross-connections exist between the auditory cortex (corpus callosum, *CC*) and inferior colliculus (commissure, *CoIC*). (Aitkin 1990, with permission Chapman & Hall) head and eyes. The EN seems to be an important link in these systems. A third system, nonspecific and alerting, concerned with attention, involves the dorsal cortex of the IC. In addition, a modality-specific set of connections supplies descending information to the brain-stem auditory nuclei.

2.4 Anatomy of Thalamocortical Auditory System

The cerebral cortex of mammals is the highest neuronal level of the brain and shows the greatest variation in phylogeny of any part of the brain. It has traditionally been subdivided into three types of cortices, each differing in the lamination patterns of constituent neurons: archicortex (including the hippocampus), paleocortex (olfactory cortex), and neocortex, considered to have evolved in that order. In mammals, each neocortical sensory area has reciprocal connections with a thalamic sensory nucleus, and it is useful to view both components as a unit. In the auditory system of the brain, afferents from the IC and efferents from the auditory cortex converge on the medial geniculate body, the auditory center of the thalamus. The most detailed descriptions of the anatomical and functional organizations of the medial geniculate body and auditory cortex are derived from studies of cats (e.g., Morest 1964, 1965; Reale and Imig 1980; Winer 1985; Morel and Imig 1987).

2.4.1 Auditory Cortical Fields

A sensory cortical field is a volume of neocortex in which the constituent neurons: (1) respond to stimulation of particular sensory receptors (cutaneous, auditory etc.); (2) are spatially organized across the cortical sheet in an orderly way in relation to the receptor distribution; and (3) collectively represent the entire receptive surface of that sensory system (skin surface, cochlea, etc.). This generally accepted definition can be attributed largely to the work of Clinton Woolsey, whose studies of the auditory system of cats (e.g., Woolsey 1960) laid the foundation for studies of auditory cortical fields in general.

Microelectrode recording techniques have been used to provide detailed maps of neuronal responses to sound in many mammals (for review, see Aitkin 1990). The frequency at which the threshold is lowest for evoking a just-detectable discharge in a neuron or multi-unit cluster is called the "characteristic frequency" (CF). Auditory cortex containing neurons with sharply defined CFs is relatively easy to map, and all mammals so studied have a "primary" auditory field (AI) across which the entire cochlea is layed out in a systematic way (cochleotopic organization). However, it may be difficult to estimate CFs of auditory neurons that are weakly responsive to sound. Such weak or sluggish responding may be an anesthetic effect – all detailed mapping studies have used anesthetized animals. Fields comprised of such neurons may show a general tendency for the base of the cochlea to be represented at one extreme of the field and the apex at the other, but

18

it may be neither a continuous nor a complete cochleotopic organization. Such fields are usually considered secondary, and there may be a gradation from highly ordered, primary-type fields to weakly ordered secondary fields.

Studies in cats (Reale and Imig 1980), mustache bats (Suga 1984), owl monkeys (Imig et al. 1977), and macaque monkeys (Merzenich and Brugge 1973) have revealed a large number of auditory fields in addition to the primary field. In cats, as many as six fields flank the AI. For other eutherians, the evidence for secondary fields is less concrete (Merzenich and Schreiner 1992). For example, in hedgehogs, considered by a number of investigators to be a primitive eutherian mammal, a primary field is flanked ventrocaudally by at least a partial representation of the cochlea (Batzri-Izraeli et al. 1990). In some studies, the design of the experiment and the relative inaccessibility of auditory tissue within sulci, along with anesthetic depression, may have contributed to the lack of evidence for secondary fields. In others, the physiologically demonstrated primary auditory field may correlate closely to anatomical indices of the entire cortical auditory representation (e.g., in quoll; Aitkin et al. 1986b; Kudo et al. 1989).

2.4.2 Connections of Cortical Fields

The history of anatomical studies of cerebral cortical connectivity is one of increasing precision in technique and the increasing combination of physiological and anatomical studies. When anatomical procedures are used, they must be related to functionally defined areas, and such areas may not always be fixed topographically in relation to gross landmarks on the cortical surface, such as sulci or blood vessels (Merzenich et al. 1975). Careful, combined use of mapping and tracing techniques, in addition to neurochemical techniques that reveal the nature of putative transmitters used by these pathways, have shown that each auditory field has a distinctive set of afferent and efferent connections, giving additional evidence for distinguishing areas in different species as equivalent. A great amount of research has been carried out on the auditory cortex of eutherians, particularly of cats and various primate species, and the reader is referred to a recent comprehensive review on auditory forebrain functional anatomy by Winer (1992).

There is thus a common plan to the auditory system in all eutherian mammals. A sound-amplification system located in the middle ear permits impedance matching between air and tissue. A transduction system in the inner ear converts, through the agency of hair cells, sound vibrations into nerve impulses. A brain auditory pathway composed of brain stem, midbrain, thalamic, and cerebral cortical regions is responsible for processing neural codes for sound, for the integration of processed neural information, and ultimately for the expression of this information as perception or acoustic reflexes.

It may be neither a continuous nor a complete crystallographic organization. Such fields are usually considered secondary, and there may be a gradation from nearly ordered prism-type fields to weakly ordered striation fields.

Chapter 3

Hearing of Marsupials

The habitats and lifestyles of marsupials are associated with diverse requirements for sensitivity to sound. We would expect a nocturnal carnivore, for example, to make more use of high-frequency transient sounds than a large herbivore. Such ecological needs are satisfied by evolutionary changes in the structure of the auditory system, brought about by selective environmental pressures (see, e.g., Manley 1972). Many writers, however, consider the hearing of modern marsupials to show primitive characteristics. For example, Frost and Masterton (1994) describe for three opossum species a poor sound sensitivity, which they suggest could be due to the attachment of the tympanic membrane to the tympanic bone in these species – a bone which is absent in eutherians. They compare their findings to the estimates by Kermack et al. (1981) of the likely sound sensitivity of the early therian *Morganucodon* and argue that opossums "have retained the primitive characteristics of still more primitive mammals".

The hearing range of a species is usually expressed in terms of sensitivity to pure tones, i.e., to sound stimuli having simple harmonic motion. However, it is unlikely that pure tones will be encountered often in nature, and it must be obvious to any amateur student of animal behaviour that most animals seem very sensitive to spectrally complex sounds, especially sounds having sharp transient changes in pressure, such as crackles and clicks. Consequently, laboratory-based studies of hearing may miss important aspects of this sense that may be more obvious to the ethologist. Nonetheless, pure-tone sensitivity remains an important index of hearing range that can be compared across species. An illustration of the likely functional significance of such sensitivity is the demonstration by Heffner and Heffner (1980) that the high frequency limit of a species' audiogram, measured using pure tones, is inversely correlated to the separation of the ears.

In contrast to postlingual humans, whose hearing can be assessed by perception and oral reporting, behavioral or physiological responses must be measured in animals and human neonates. The sensitivity to sound of a species is usually described by measuring an *audiogram*, in which the threshold sound level needed for a species to hear a particular pure tone is plotted against the frequency of the tone. The response indicative of hearing can take the form of an unconditioned or conditioned reflex, or of a physiological response, such as a sound-evoked potential.

3.1 Behavioral Measurements

Ehret (1983a) has reviewed the literature on the various *unconditioned motor responses* employed as indices of hearing. Startle responses – limb or body twitches, usually evoked by loud sounds – are not considered by most authors to be sensitive indicators of auditory thresholds; such responses are, however, useful qualitative indices of hearing. If the animal responds, it must have heard. The Preyer reflex is a twitch of the pinna which can be evoked by relatively soft sounds. Orientation responses include a component of body orientation towards the sound, and various autonomic responses (heart rate increases, skin resistance changes) can also be evoked by sounds. Some use has been made of these responses to examine the relationship between frequency and threshold intensity for evoking the response (e.g., in neonatal kittens and mice; Ehret 1983a).

Of more value are the various conditioned responses, and the reader is referred to the compendium of animal psychophysical material collected by Fay (1988). Two forms of conditioning are commonly used – aversive and operant. Aversive conditioning requires that the animal carry out a motor task in order to avoid a shock, whereas in operant conditioning the motor response procures a reward. In both cases a sound serves as a warning signal. The animal is trained to respond to a certain sound frequency of relatively high intensity; the latter is then diminished until response rate reaches chance level, at which intensity the threshold to that sound frequency is considered to have been reached.

One technical problem with the investigation of hearing ranges of animal species, absent from the assessment of a human's audiogram, is the uncertainty that the animal is responding to the principal frequency being presented. Although sound systems can be calibrated accurately for those frequencies intentionally presented, the extent to which animals respond to other frequencies, generated by the spectral "splatter" which occurs when tone pulses of rapid rise time are applied to a loudspeaker, may be difficult to gauge. The problem is minor when frequencies audible to humans are used but becomes more serious the higher the ultrasonic frequency. Is the animal responding to that particular ultrasonic frequency, to some subharmonic closer to the center of the animal's hearing range, or even to a distortion product generated in the cochlea and not detectable from the sound system?

Behavioral audiograms have been measured using aversive conditioning for three marsupial species (Fig. 3.1). Ravizza and his colleagues (1969) used a conditioned suppression paradigm in their study of the opossum, *D. virginiana*. This species has a rather shallow insensitive audiogram with a best frequency of hearing between 16–32 kHz at 20 dB SPL. Two other opossums, *M. domestica* and *Marmosa elegans,* studied using the same technique, have similarly flat, insensitive curves (Frost and Masterton 1994) which are shifted towards higher frequencies. These curves have a high-frequency upper limit similar to many eutherians (e.g., rat, Fig. 3.1) and cats (Heffner and Heffner 1985a). On the basis of this and the evidence that the bone structures of modern opossums are very similar to fossil material from Cretaceous times, Frost and Masterton (1994) have argued that sensitivity to high frequencies was present in ancient mammals.

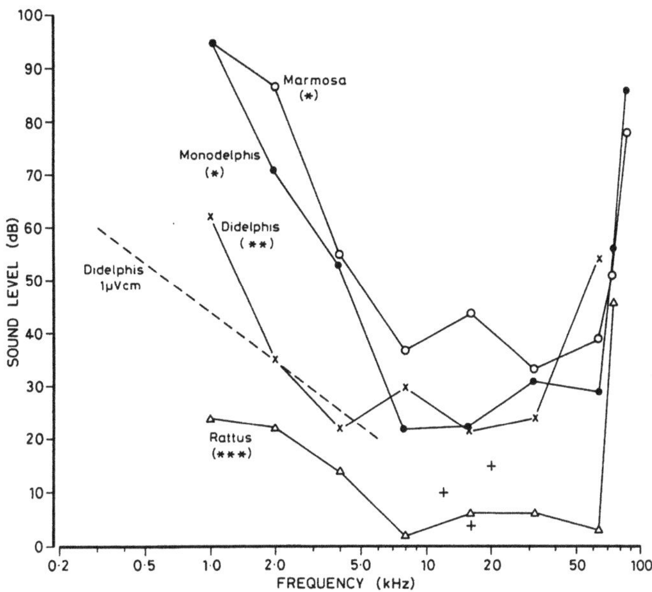

Fig. 3.1. Behavioral audiograms of three opossums and *Rattus rattus* from the studies of *Frost and Masterton (1994); **Ravizza et al. (1969); and ***Kelly and Masterton (1977) (data kindly supplied by the late Dr. R.B. Masterton). The *plus symbols* (+) are behavioral thresholds for *Monodelphis* from the study of Reimer and Baumann (1995). The *dashed line* is the slope of the function relating the sound level needed to evoke a one microvolt cochlear microphonic potential as a function of tone frequency (Redrawn from McCrady et al. 1937)

Reimer and Baumann (1995) used a simple operant-type paradigm that utilized the tendency of the *Monodelphis* opossum to eat small amounts of food at any one time. A tone was presented from one of two loudspeakers located at the end of two arms of a Y maze. Food dispensers were located at the base of the sound sources, and the animal received a food reward if it chose the correct loudspeaker. The loudspeaker receiving the pure tone was randomized, and no more than four successive trials were initiated from a speaker.

The resulting audiogram, assessed using a similar range of test frequencies to those used by Frost and Masterton (1994), differed in one main respect from that study – thresholds. Those in the mid-range were more sensitive in the study of Reimer and Baumann; their threshold of 4 dB SPL at 16 kHz is more like that of eutherians of equivalent size (Fig. 3.1, "+"). Reimer and Baumann suggested that the difference between the two studies could be due to motivation (better with positive than negative reinforcement) and, possibly, age (young animals in the Reimer and Baumann study; unspecified age in that of Frost and Masterton).

These two studies indicate that training methods may have an important influence on the results of behavioral studies. Other supporting evidence may be needed to confirm audiograms measured using one procedure. In this respect Reimer (1995) has measured auditory evoked potentials (see below) in *Monodelphis* opossums, obtaining an audiogram with a similar shape and sensitivity to that obtained in her study with Baumann (1995). The sensitivity of the audiogram in the mid-range of frequencies diminished by about 15 dB between animals of 3 and 4 months of age to achieve a value similar to that reported by Frost and Masterton (1994).

3.2 Electrophysiological Measures

Behavioral training of animals, especially wild animals, is extremely time consuming. A much simpler approach is the measurement of electrical activity from the auditory nervous system, from cochlea to auditory cortex. This requires that the animal be anesthetized, if only, as in the case of noninvasive techniques such as scalp recording, to eliminate movement. Although anesthesia is likely to reduce the excitability of the nervous system, so that thresholds for neural responses may be higher than behavioral responses, it would be expected that threshold elevations will apply to all frequencies and not be selective. Consequently, the *shapes* of electrophysiological and behavioral audiograms should be quite similar.

It is usually taken for granted by experimenters that anesthetic agents will have similar effects on all mammals, provided a correct dosage regimen is established for each species. However, in the present author's wide experience with marsupials, the same anesthetic can have widely variable effects between species, even within a single species (e.g., Aitkin et al. 1979), and commonly used agents, such as pentobarbitone sodium, may sometimes be ineffective or lethal in the same species and dose rate. As a consequence, prolonged procedures and data collection possible under anesthesia in laboratory species, such as cats and monkeys, have rarely been achieved with marsupials. Marsupials would thus not appear to be animals of choice for long-term experimentation under anesthesia.

3.2.1 Cochlear Potential Thresholds

The process of recording cochlear potentials from the round window of marsupials is hampered by the complex anatomy of the skull and middle ear in these species. In eutherian mammals there is a fusion of certain bone elements into the hollow bulla and temporal bone. This does not occur in opossums (McCrady et al. 1937). The tympanic bone retains its primitive annular shape, the endoderm of the external auditory meatus forms part of the wall of the middle-ear cavity, and there is no bulla. The middle ear is thus separated only to a slight degree from adjacent tissues by bony projections. This feature is observed in many other marsupials, both polyprotodontid and diprotodontid (unpubl. observ.). The net result is difficulty in exposing the round window of the cochlea for the placement of electrodes or perfusion of the cochlea.

In adult opossums the mastoid portion of the squamosal bone overhangs the round window in such a way as to hide it from a direct lateral view. The problem is less apparent in younger animals, because the processes of the squamosal bone are smaller; at ages younger than 100 days there is no obstacle at all to a direct lateral view. The cochlear microphonic audiograms recorded from these young opossums were among the first measured in any species (both *D. virginiana* and *D. marsupialis*) by McCrady et al. (1937, 1940). These workers do not illustrate CM "thresholds" above 10 kHz, so their results are not directly comparable with those of Ravizza et al. (1969). However, they do indicate responses up to 25 kHz, relatively small responses at a given intensity compared with the eutherian guinea pig, and a "striking lack of sensitivity" to low frequency tones; similar features were observed in both species. A line of best fit drawn through the low-frequency points for the 77-day-old *D. virginiana* illustrated by McCrady et al. (1940) has a much more gradual slope than the low-frequency arm of the behavioral audiogram measured by Ravizza and colleagues (1969; Fig. 3.1).

Recordings have also been made from the round window of the brushtail possum (Aitkin et al. 1979). In adults of this species, the round window is more accessible than in opossums and is covered laterally by a spongy bone which can be removed without interference with the auditory system. The best frequency of the 50-μV CM curve in brushtail possums is skewed towards low frequencies (below 1 kHz); subsequent work (see below) suggests that these data have little bearing on the hearing range of the brushtail possum. The same consideration may apply to the CM "audiogram" published for platypus (Gates et al. 1974), although the relative insensitivity and restricted frequency range of the platypus CM curve do parallel another measure of hearing sensitivity, the function relating stapes movement amplitude to frequency, in one of the two species (*Tachyglossus*) of the only other monotreme, the echidna (Aitkin and Johnstone 1972).

Measurements of the first peak in the summed auditory nerve response (N_1) from the round window or within the cochlea are more useful for constructing neural audiograms, because they are more selective, and the origins of the potentials are better understood (Johnstone et al. 1979; Rajan et al. 1991). Averaged responses to brief, fast rise time, low-intensity tone bursts originate from auditory nerve fibers supplying the region of the organ of Corti, maximally influenced by the particular frequency used. Audiograms measured using this technique (e.g., cat, Rajan et al. 1991) differ from behavioral curves (e.g., cat, Heffner and Heffner 1985) in absolute sensitivity and upper-frequency limit, but are similar in other respects. Up to now the technique has not been used with marsupials.

Audiograms may also be determined by measuring the thresholds for distortion products generated by cochlear mechanisms as a function of frequency (Gaskill and Brown 1990). The audiogram so obtained for *Monodelphis* (Faulstich et al. 1996) resembles the shape of the behavioral audiogram measured by Frost and Masterton (1994) but lies close in threshold sensitivity to those of Reimer (Reimer 1995; Reimer and Baumann 1995).

3.2.2 Single-Unit Thresholds

Measurements of the threshold sound pressure level required to activate single neuronal cell bodies or axons (collectively called "units") in the auditory pathway (see Aitkin 1990 for details) has been used to estimate the hearing range of an animal. When the thresholds at characteristic frequency of many units sampled throughout an auditory nucleus or the auditory nerve are plotted against CF, the scattergram parallels the behavioral hearing curve of the animal with the latter being 10–20 dB more sensitive at the highest frequencies (Konishi 1970; Evans 1972; Dallos et al. 1978). The difference between the behavioral and electrophysiological audiograms may lie in the way stimuli reach the ear in the two types of experiment. Most electrophysiological experiments utilize *dichotic* stimulation, in which precisely controlled stimuli are presented as closely as possible to the eardrum, bypassing the pinna. Behavioral experiments, in contrast, require *free-field* stimulation – sound reaches the eardrum after being amplified by the pinna. The thresholds in free-field electrophysiological experiments are likely to be much lower than those in experiments using dichotic stimulation (in which stimuli are presented by earphones or by tubes inserted into the meatuses) (e.g., Moore et al. 1984). However, single-unit CF thresholds from the cochlear nucleus of *Monodelphis* (Müller et al. 1993) are up to 20 dB *more* sensitive than behavioral curves (Frost and Masterton 1994; Reimer and Baumann 1995).

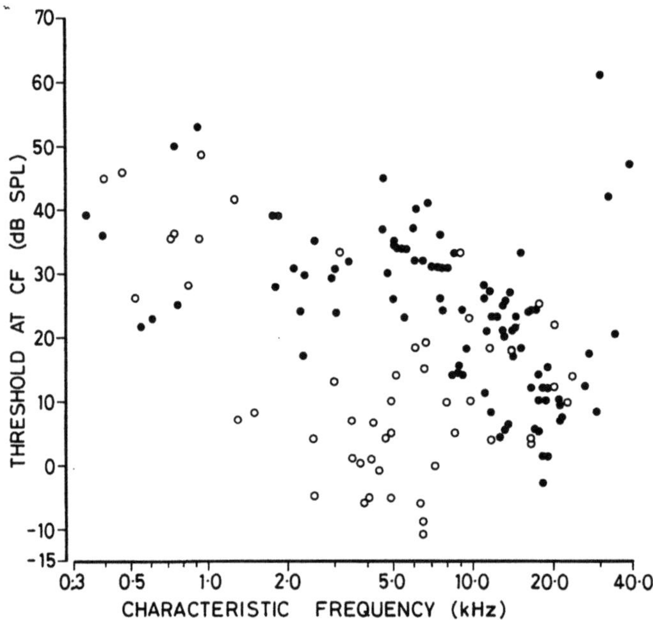

Fig. 3.2. Thresholds in decibels in relation to 20 µPa (*dB* sound-pressure level [*SPL*]) for the just-detectable discharge of single units in the inferior colliculus of brushtail possums, recorded under either dichotic (*filled circles*) or free-field (*open circles*) stimulation (see text). (Data replotted from Aitkin et al. 1978a, 1984)

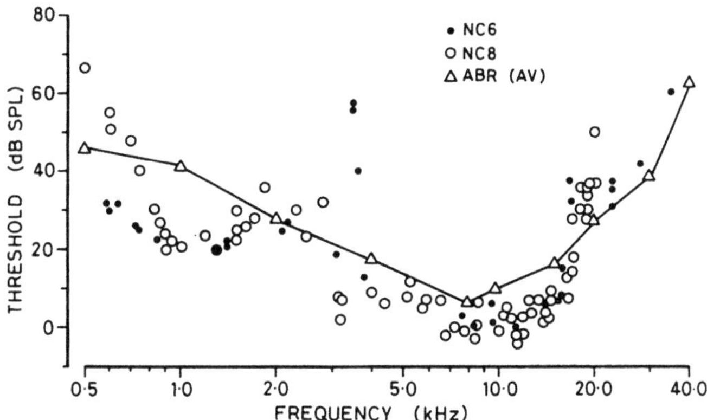

Fig. 3.3. The average ABR audiogram (*open triangles*) of four female Northern quolls compared with the unit and unit-cluster thresholds from the inferior colliculus of two Northern quolls (*NC6* and *8*). (Aitkin et al. 1994b, with permission of Wiley-Liss, Inc.)

Among marsupials, single-unit data have been collected in brushtail possums (Fig. 3.2) and Northern quolls, also known as native cats (Fig. 3.3). The scattergram of single-unit thresholds from the auditory cortex of brushtail possums (Fig. 3.2, filled circles), obtained with dichotic stimulation (Gates and Aitkin 1982), reveals a most sensitive frequency for hearing near 20 kHz. In experiments on the inferior colliculus using free-field stimulation (Aitkin et al. 1984), the frequency-selective amplifying effects of the pinna (see Chap. 5) appear as a substantial reduction in single-unit CF thresholds in the mid-range, 1–10 kHz (open circles, Fig. 3.2). Given the variability in these data due to factors such as anaesthetic level, the best frequency of hearing in the brushtail possum is likely to lie somewhere between 7–20 kHz.

The scattergram of single-unit thresholds from the inferior colliculus of two quolls (Northern native cats, NC; Fig. 3.3), obtained in experiments using dichotic stimulation, has a minimum between 7–12 kHz and lacks a further peak at 20 kHz. There is a 10–20 dB rise in threshold between 1.5–3 kHz, not present in the ABR audiogram (curve in Fig. 3.3; see below), which may be explicable in terms of pinna amplification.

3.2.3 Auditory Brain Stem Response (ABR)

With the aid of computer averaging, brainstem potentials evoked by transient acoustic stimuli can be recorded by low-impedance electrodes attached to the scalp. The resultant waveform, the auditory brainstem response (ABR), consists of six or more positive deflections, each of which can be related to activity in an auditory nucleus or tract. Thus, waves I and II are believed to originate from activity in the cochlear nerve, and wave IV is usually identified with the midbrain auditory nucleus, the inferior colliculus (Jewett 1970).

27

An ABR "audiogram" is determined by presenting brief pure tones with fast rise times and averaging the responses to hundreds of such tone bursts. The amplitude of the ABR will be related to the number of neurons activated, and this in turn will be determined by the frequency and intensity of the tone burst. The visual detection of a particular peak (commonly wave IV) can be used to estimate the threshold at that frequency. The absolute amplitudes of waves in the ABR and their sizes in relation to the background noise level will influence the usefulness of this technique. In order to diminish the amount of muscle electrical activity, animals need to be anesthetized. The anesthetic is likely to reduce the excitability of the brain-stem neuronal generators, thus giving higher thresholds, but is unlikely to affect latency or amplitude/intensity functions (Bobbin et al. 1979; Gossampson and Kirss 1991).

The ABR has been used as an index of human hearing for many years, and clinical data suggest that there is a good agreement between frequency specific ABRs and pure-tone thresholds in the majority of patients (Pratt and Sohmer 1978; Smith and Simmons 1982; Davis et al. 1985; Fjermedal and Laukli 1989). Published correlations for humans are usually only available up to 8 kHz. Studies in animals suggest that ABR audiograms and behavioral threshold curves are generally similar in shape, but that ABR thresholds are higher than behavioral thresholds, especially at high frequencies (Wiener et al. 1966; Hood et al. 1991; Xu et al. 1993). Thus, ABR audiograms are useful indicators of a species' hearing range, if not its absolute sensitivity.

Measurements of ABR thresholds have been made in a number of marsupial species: *Monodelphis* opossums (Reimer 1995); tammar wallaby (Liu et al. 1996; Cone-Wesson et al. 1997); three arboreal polyprotodontids (unpubl. observ.); and three members of the family Dasyuridae (Aitkin et al. 1994b; Fig. 3.4). The Northern quoll (*Dasyurus hallucatus*) is a nocturnal carnivore from the tropical north of Australia and weighs about 300 g. The kowari (*Dasyuroides byrnei*) and stripe-faced dunnart (*Sminthopsis macroura*) live in arid areas of central Australia; they weigh about 120 and 27 g, respectively. The ABR audiograms for the three species have minima at 8 kHz (quoll and kowari) or 10 kHz (dunnart). The shapes of the audiograms are quite similar, with that for the dunnart displaced along the frequency axis to the right, having poorer low-frequency thresholds and a 20 dB better threshold at 40 kHz.

The tone thresholds for ABRs of the tammar wallaby are highly variable from one subject to another (Cone-Wesson et al. 1997). This may be a result of the large head size of the wallaby, with the resulting large distance and possible nonauditory electrical interference between the current sources in the brain responsible for the ABR and the recording electrodes. However, the lowest thresholds recorded were near 0 dB SPL and occurred between 8–16 kHz, a similar frequency range to that in the smaller marsupials named above and the domestic cat (Heffner and Heffner 1985a).

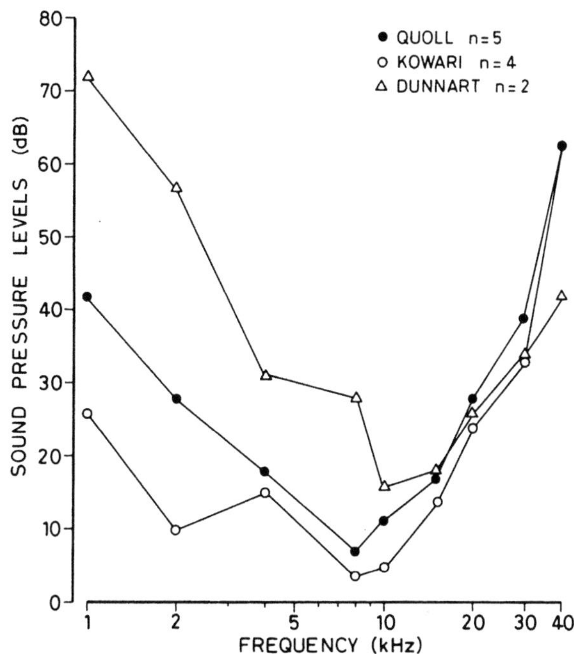

Fig. 3.4. ABR audiograms of three species of Dasyuridae: average of five Northern quolls (*filled circles*), four kowaris (*open circles*), and two dunnarts (*triangles*). (Unpubl. observ. of L. Aitkin, J. Nelson, and R. Shepherd)

3.3 Future Research Directions

With a few exceptions, such as *Monodelphis*, marsupials are not laboratory animals, and the hearing ranges of only a few wild marsupials have been studied. Some, such as the kangaroos, are very common in nature in Australia. Of these "wild" marsupials, much of the published information has been obtained from animals held in colonies in laboratories, and it is difficult to know how the hearing of these species, often reared over many generations in buildings, in the absence of natural environmental sounds and in the presence of potentially damaging human-made noises, relates to that of creatures in the wild. There are several factors in addition to the species itself that will determine the shape and sensitivity of an audiogram – the method of testing, the sex and age of the subject, and the acoustical system generating the test sounds, for example.

Measurements of hearing in field studies using portable devices such as those for measuring ABRs could provide illuminating data, because the subjects would be tested in a natural environment. Such studies will be fraught with their own limitations but can be used with minimal disturbance to the animal in its own environment. It will be interesting to see if marsupials fit in to the general relationships, observed for mammals by Heffner and Heffner (1980, 1992), between hearing range, directional acuity, and other anatomical and sensory features.

Chapter 4

What Do Marsupials Listen To?

The auditory pathway is one of three sensory systems in eutherians which permit detection of events occurring at some distance from the receiver. The relative importance of vision, audition, and olfaction for a species depends on both ecological and phylogenetic factors. For example, diurnal animals are likely to make much use of vision, but nocturnal animals and those active during the day in a forest canopy will have a more restricted visual input and may depend more on auditory input to evaluate important events occurring in the environment. Our examination of the importance of hearing to marsupials must take into account the vocalizations made by conspecifics, both adult and neonatal, and the acoustic stimuli generated by the movement of prey and predators. The acoustic spectra of these sonic events need to be examined in relation to the abiotic sounds generated in the acoustic biotope of each species.

4.1 Vocalizations and Speech

Most vertebrate species vocalize, and most of these vocalizations are innate, identifying such major behavioral states as fear, hunger, rage, and sexual drive. For a given animal, a vocalization with a certain spectral composition will always mean the same thing to members of that species (conspecifics). This also applies to the cries of human babies.

Human speech, however, must be learned. Deaf children have very limited speech capabilities even though they may have normal vocal tracts. The phonemes of human speech do not have a fixed meaning, but change their significance depending on the phonetic context. For example, the vowel /a/ has a different "meaning" in the contexts m/a/n, m/a/ and /a/n. Some nonhuman species have elaborate vocalizations that also require learning from an adult – many songbirds, for example (Marler 1970), and some primates (e.g., vervet monkeys; Struhsaker 1967; Cheney and Seyfarth 1982; Seyfarth and Cheney 1986) – so that learning per se is not unique to human speech.

There are limitations in the *abilities* of nonhuman species to generate vocalizations. These arise because of the differences in the structures and innervation of the mouth and larynx in humans and other mammals (see Ploog 1988). In humans, the increased mobility of the tongue and lips, and the increased angulation between the mouth and the upper respiratory tract facilitate the generation of a vast array of possible speech sounds. According to Lieberman (1984), the ability to form extreme vowel positions occurs only in humans and only with development. In animals, voicing is due to vocal cords, whereas human speech is due mainly to articulatory movements of the tongue, velum, and lips. In passing, it is interesting to note that amphibians, birds, and humans all use respiratory air flow to produce communication sounds; however, their vocal tracts operate differently (Green and Marler 1979).

There are parallel differences in the motor pathways for nonspeech and speech vocalizations. In macaque monkeys, for example, vocalization is initiated by activity in the limbic system and thalamus (Jürgens 1988). For humans, the integrity of neocortical regions, such as Broca's area and the supplementary motor area, is essential for speech. Damage to the human motor face area leads to anarthria and aphonia. In monkeys, such damage has little effect on vocalization because, as mentioned, such vocalizations relate mainly to vibrations of the vocal cords and not the face or lips.

4.2 Vocal Behavior of Adult Marsupials

Marsupials are not a particularly vocal group of animals, although the circumstances in which vocalizations are used may not often be encountered by the casual observer. Most marsupials have nocturnal activity patterns, and most of their vocalizations are directed to conspecifics, usually during mating, territorial, and mother/young encounters. Rarely do marsupials vocalize in response to humans, except in circumstances in which fear is aroused, so that isolated captive animals provide little vocal information. As a consequence, high-quality recordings suitable for spectral analyses are available for only a few species.

Characterizations of vocalizations have traditionally used onomatopoeic descriptions – "hiss," "click," "churr," etc. These are often backed up by measurements of call spectra, sonagrams and spectrograms, with the quality of these dependent on the recording and analytic systems, especially the upper limits of their frequency response. On the basis of careful physical measurements of the vocalizations of seven dasyurids and six didelphids, Eisenberg et al. (1975) suggested that marsupial vocalizations belong to four categories: (I) tonal, with energy in narrow frequency bands ("squeak, chirp, moan"); (II) noisy sounds with tonal components ("growl, churr"); (III) clicks, of short duration and of little harmonic structure; and (IV) noises with a broad frequency spectrum ("hisses, screams"). This scheme has been extended to other species (Croft 1982; Biggins 1984). Loose associations between behavior and vocalization suggested that type I sounds related to low levels of adult arousal, such as mother/neonate communication, type II with moderate levels of arousal, type III with close physical associations, investigative without overt arousal, and type IV sounds with high arousal and threatening situations.

The calls of adult opossums – "hiss," "growl," and "screech" – are all given in agonistic or defensive situations and have very broad spectra (0.5–8 kHz, growl; 1–16 kHz, screech; McManus 1970). Most of the calls of dasyurids have a dominance of frequencies below 3 kHz in all but the smallest species (*Sminthopsis*; Eisenberg et al. 1975). Although ultrasonic frequencies can be recorded, they do not contribute much energy to the overall signal. More recent work on *Dasyurus hallucatus* (Northern quoll) by Dempster (1994) found little evidence for calls with harmonic structure or frequency modulation; all vocalizations were broadband, noisy sounds comparable to type IV above.

Power spectra of the "hiss" call, a type IV sound made by an adult Northern quoll during intense interspecific arousal, has most of its energy below 2 kHz (Fig. 4.1). An equivalent call made by a juvenile Northern quoll (109 days old) would be classified as type II since it has peaks of energy at 6–7 kHz and 12 kHz (Fig. 4.1). These higher frequency peaks are probably the remnants of the spectra of the isolation calls made by dependent pouch young of this species (Aitkin et al. 1994b, 1996a). The vocalization of a smaller dasyurid, the kowari, obtained during a male-female social interaction, has its peak energy (Fig. 4.1) near the best frequency of hearing of this species (Fig. 3.4). There is only 15 dB variation in sound pressure between 0.8 and 20 kHz, but a clear peak is present at around 8 kHz. Although this call could be classified as type IV, the relatively nonthreatening social situation in which it was recorded suggests type II. Finally, the cry of a striped possum, an arboreal marsupial, obtained during moderate arousal, has clear tonal peaks at 2.5 and 5.5 kHz, superimposed on a broad frequency range, and so would be classified as type II (Fig. 4.1).

Arboreal marsupials generally seem to be more vocal than terrestrial marsupials, although much of the information is based on onomatopoeic description rather than measurements (Biggins 1984). The arboreal habitat and nocturnal behavior have meant that recorded vocalizations are often difficult to relate to social behavior or to the sex of the emitter. Recordings have been published for the yellow-bellied glider (*Petaurus australis*; Kavanagh and Rohan-Jones 1982), brushtail possum (*Trichosurus vulpecula*; Winter 1977) and koala (*Phascolarctos cinereus*; Smith 1980), and the reader is referred to a summary of calls of various

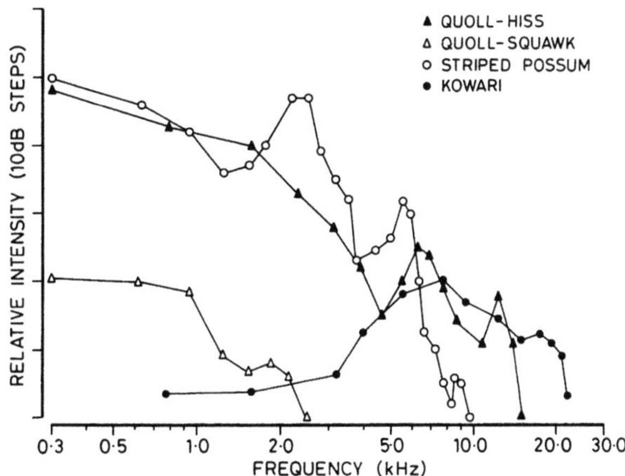

Fig. 4.1. Power spectra of four calls plotted as relative energy at a given frequency. The ordinate applies to any given call, but intensities cannot be compared between calls. The frequency response of the recording system was linear to 15 kHz for the striped possum and to 22.5 kHz for the other three calls

33

possums and gliders expressed onomatopoeically in relation to the social contexts in which the sounds normally occur (Biggins 1984).

4.3 Hearing Sensitivity and Vocal Spectra of Adults

Konishi (1970) showed that single-unit thresholds correlated closely with behavioral audibility curves in songbirds and that the upper range of frequencies, although varying between species, correlated with the differences in the range of vocal frequencies. In most songbird species the dominant vocal frequency does not correspond precisely to the best frequency of hearing but lies above that frequency; furthermore, the audibility curve extends below the range of frequencies present in the vocalization. Clearly, audiogram "space" is also available for other sounds, for example, those of predators and of conspecific chicks.

The relationship between hearing range and vocalization frequencies has not been studied in detail in mammals, but a recent study of the Northern quoll has reviewed some of the literature (Aitkin et al. 1994b). The vocalizations of adult quolls fall into three major groups – "hiss," "squawk," and "sniff" – and some less common sounds, including a twittering sound containing ultrasonics, given both by solitary animals and during interactions (Dempster 1994). The power spectra of the major calls show a concentration of energy in the lower part of the quoll

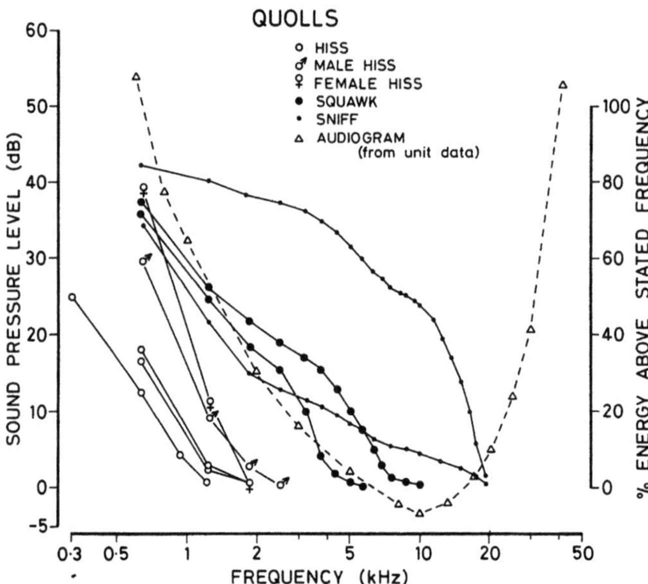

Fig. 4.2. Relationship between call spectra and hearing range in a carnivorous marsupial. Power spectra of selected calls of the Northern quoll plotted as the percentage of energy above the stated frequency (*right ordinate*), compared with the audiogram from unit threshold data from the study of Aitkin et al. (1986b), to which the *left ordinate* (*dB*) relates. Upper frequency limit of the recording system was 22.5 kHz

34

audiogram; only the "sniff" call extends into the central part of the audiogram (Fig. 4.2).

The behavioral audiogram of the opossum (*Didelphis virginiana*) demonstrates a best frequency of hearing between 16–32 kHz (Ravizza et al. 1969), at least an octave higher than the quoll. First contact between adult conspecifics evokes a hiss, which deepens in pitch and becomes higher in amplitude as the disturbance continues, becoming a growl (McManus 1970). The latter has a spectrum between 0.5–8 kHz. More extreme disturbances evoke screeches which include components at 2, 5, 10, and 15 kHz. Both of these major call types have energy falling mostly in the lower half of the adult hearing range.

Brushtail possums have a variety of broad spectrum vocalizations including a characteristic rumbling hiss (Winter 1976). Analysis of some of these calls indicate that they are dominated by frequencies below 4 kHz (unpubl. observ.) and would relate mainly to the low frequency arm of the audiogram. The presence of ultrasonic components cannot be precluded since the analysis system had an upper frequency limit of 12 kHz, but there was little or no energy in recorded calls between 4–12 kHz (unpubl. observ.).

The ABR audiograms and vocal spectra of two arboreal marsupials are shown in Fig. 4.3. Feathertail gliders (*Acrobates pygmaeus*) make "ticking" or "popping" sounds (type III of Eisenberg et al. 1975); in other marsupials these sounds are made by adult males investigating estrus females (Biggins 1984). Most of the energy of these feathertail glider calls lies below 8 kHz; one such call recorded from a male-female pair may be compared with the average ABR audiogram measured in both members of this pair (Fig. 4.3A). Although generally rather insensitive, the ABR audiogram of this species appears to have two regions of relatively low threshold, and the spectra of the "click" calls correspond well with the lower frequency range of the audiogram.

Ringtail possums (*Pseudocheirus peregrinus*) are gregarious, moving in small groups through trees in search of leaves and flowers. The young, after leaving the pouch, make brief chirping contact sounds when accompanying their mothers. The average ABR audiogram of an adult male ringtail possum has a minimum at 4 kHz with a threshold of 6 dB (Fig. 4.3B). The isolation calls of a juvenile ringtail possum, recorded elsewhere, have spectral peaks at 4–4.8 kHz, in very close register with the best frequency of hearing of the adult.

There is vast literature on the hearing and vocalizations of humans, but parallel studies of hearing range and vocalization spectra are rare in both eutherians and marsupials. There are intriguing studies of elephants: one can correlate the low best frequency of hearing in an Indian elephant (best at 1000 Hz, 5-dB threshold, but still about 40 dB at 30 Hz; Heffner and Heffner 1980) with the very low frequency (sometimes infrasonic) calls made by African elephants (Poole et al. 1988). More extensive information, still of a correlative nature, exists about laboratory species. The spectra of most calls of the adult cat contain frequencies well below the best frequency for hearing (Heffner and Heffner 1985a; Aitkin et al. 1994b). The ferret (*Mustela putorius*) has a broad hearing range extending from 36 Hz–44 kHz at 60 dB and a best frequency of hearing at 8–12 kHz (Kelly et al. 1986). Sonagrams for hiss, pant, chatter, and scream calls of a related species, the Siberian ferret (*Mustela eversmanni*; Farley et al. 1987) all have dominant fre-

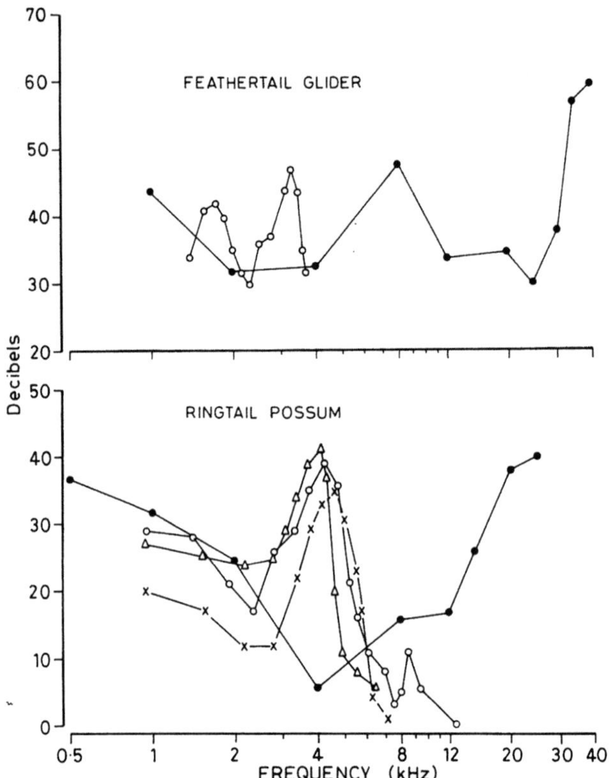

Fig. 4.3. Relationship between call spectra and hearing range of two arboreal marsupials assessed by thresholds of ABRs to tone pips. Feathertail glider (*filled circles*): average ABR audiogram of 3 animals (1 male, 2 females); the "click" sounds (*open circles*) evoked by one of a male–female pair held captive in a small cloth bag. (Animals kindly supplied by Dr. Simon Ward.) Ringtail possum (*filled circles*): ABR audiogram from a single male; the three juvenile isolation calls (other symbols) were made by other ringtail possums in a separate study in a woodland environment. (Data kindly supplied by N. Linahan.) Frequency response of the recording system linear to 22.5 kHz (feathertail) or 15 kHz (ringtail)

quencies (2–5 kHz) well below the best frequency for hearing of *M. putorius*. The least weasel (*Mustela nivalis*) hears from 51 Hz to 60.5 kHz at 60 dB with a rather broad region of best sensitivity from 1–16 kHz (Heffner and Heffner 1985b). The vocalizations of this smaller mustelid, although generally higher in frequency than those of the ferret (1.5–8 kHz), are still restricted to the lower part of the hearing range of this species (Huff and Price 1968). These authors tested for, but failed to find, ultrasonic sounds from an adult female and her four 3-week-old young.

It may be, for nocturnal carnivorous species such as cats and mustelids, that the major part of the hearing range – best frequency and high frequency arm of the audiogram – are "available" for detecting the sounds of prey (or predators, in the case of the feathertail glider). Examination of nonpredatory and herbivorous species may give some insight into this speculation. The audiogram of the adult

36

laboratory rat has sensitive minima at 8 and 32 kHz, and has a range from 4–54 kHz at 10 dB above threshold (Kelly and Masterton 1977). The audiogram of the house mouse extends from 0.5–120 kHz. It has a minimum at 15 kHz, with sensitivity persisting above 32 kHz (Heffner and Masterton 1980); other strains of mice have an additional peak at 50 kHz (Ehret 1983b). Unlike those of nocturnal carnivores, such as quolls, the spectra of the vocalizations of these laboratory rodents seem to extend across their ranges of hearing (for review, see Aitkin et al. 1994b). Some of the sounds made by predators, such as cats and owls, are likely to be audible to rodents, but they must also use other strategies to avoid predation.

Audiograms have been measured for at least 18 species of subhuman primate (Fay 1988), all of which are primarily vegetarian, and there is a substantial body of literature on primate vocalizations (e.g., Todt et al. 1988). Some of the arboreal species, such as the squirrel monkey and marmoset, are highly vocal; examination of their calls may cast some light on what might be expected of the relation between hearing and vocal communication in arboreal marsupials. A comparison of the power spectra of three types of marmoset calls, measured in my laboratory, with a behavioral audiogram augmented by single-unit threshold from experiments in my laboratory (Seiden 1957; Aitkin and Park 1993), reveals strong correlation between audiogram sensitivity and dominant frequencies in the vocalizations. The "cough" sound, uttered during arousal by an aggressive animal (Epple 1968), has most of its energy at low frequencies, below 4 kHz (Fig. 4.4). The energy of the isolation, "phee", call is concentrated between 6–8 kHz, and that of the "twitter" call, made in isolation but with loose visual contact (Epple 1968), is around 15 kHz (Fig. 4.4). The dominant frequency ranges of these sounds do not overlap significantly (note that the reduced hearing sensitivity between 3–5 kHz is associated with low energy in all the calls shown here), but collectively span the entire hearing range of common marmosets. Perhaps for socially gregarious

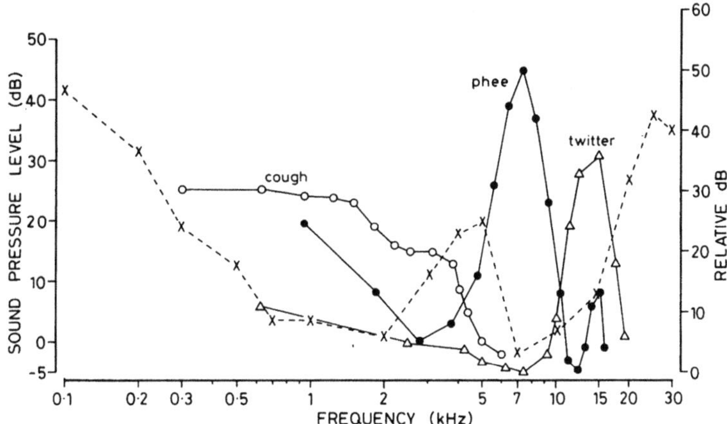

Fig. 4.4. Relationship between call spectra and hearing range of an arboreal primate. Three well-described calls of the common marmoset (see Epple 1968) are compared with the behavioral audiogram of this species (Seiden 1957; Fay 1988). The calls were recorded with an upper frequency limit of 22.5 kHz

species, such as marmosets and some of the gliders and possums (McKay 1989), intraspecific vocal communication is the major role of the auditory system. This hypothesis is amenable to testing with other arboreal species.

4.4 Adult and Neonatal Hearing and Vocal Relationships

The development of hearing and vocalizations have been studied in detail in Northern quolls (Aitkin et al. 1994b, 1996a). The sounds made by the mother quoll when tending her young have not been recorded; anecdotal observation suggests few sounds are made in the presence of human observers. The ABR audiograms of young quolls are first measurable when hearing begins, at 60–65 days, and have a similar U-shape to those of the adult, with the best frequency lowered by about an octave to 4kHz (see Chap. 8). The auditory sensitivity conveyed by such hearing ranges would not be particularly suitable to most common adult quoll calls (Dempster 1994), and prior to day 60 the young quoll is unlikely to hear anything at all. In relation to juvenile hearing, marsupials such as quolls are likely to be highly altricial compared with avians.

However, the mother quoll appears to be very sensitive to the sounds made by the young. Pouch-young quolls do not spontaneously vocalize until about 35–45 days when they begin to move from one nipple to another. The calls emitted vary somewhat in time structure (Aitkin et al. 1996a); some have an abrupt onset (e.g., 36, 52 days), others a slower onset (45 days), or both slow and fast components (65 days). Call durations are mostly 100–200 ms, but pouch young also emit very short calls (85 days) and "click" sounds. There is no obvious difference between the waveforms (or the spectra) of male or female pouch young. In older pouch young and adults, the common "hiss" sound (e.g., 109 days) lasts 1–2 s in adults, although some of the vocalizations of older animals have durations of 200–300 ms (Dempster 1994).

Fourier analyses of the waveforms of calls elicited by animals 65 days or younger reveal a concentration of energy between 2–20 kHz, with peak intensity of the isolation call around 10 kHz (Fig. 4.5). The spectrum of the call changes abruptly between 65 and 85 days, with a new peak developing near 6 kHz and a plateau between 100 Hz–1000 Hz. Beyond this age (e.g., 109 days; Fig. 4.5), output is greatly attenuated at high frequencies, and most energy in the call lies below 2000 Hz, as in adults (Aitkin et al. 1994b; Dempster 1994). For quolls younger than 65 days, the frequencies at which peak energy occurs cluster closely about the best frequency of adult hearing (Aitkin et al. 1994b).

A similar match occurs between the peak energy of isolation sounds made by neonatal *Monodelphis* and the behavioral audiogram of the adult (Figs. 4.6, 4.7). The peak energy is concentrated around 10–13 kHz. With some calls, a second component is present at a higher frequency which may be an octave above the first component (Fig. 4.6). The most sensitive part of the hearing range lies between 8–16 kHz, although thresholds at 32 and 64 kHz are only about 10 dB less sensitive (Frost and Masterton 1994), so that the second component also falls in a sensitive region of the hearing range of *Monodelphis*.

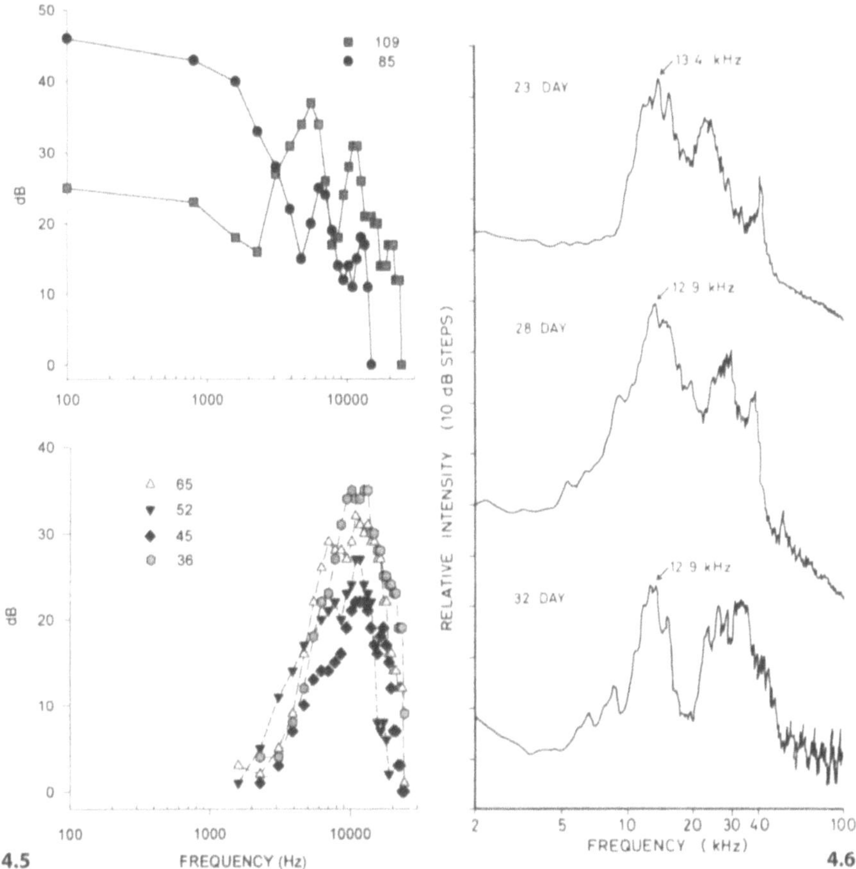

Fig. 4.5. Power spectra of vocalizations made by quoll pouch young. Note the concentration of energy around 10 kHz for animals younger than 65 days, and the extension of energy to lower frequencies at 85 and 109 days, by which time the quoll is weaned. Frequency response of the recording system linear to 22.5 kHz. (Aitkin et al. 1996a, with permission of Wiley-Liss, Inc.)

Fig. 4.6. Power spectra of vocalizations made by neonatal *Monodelphis* opossums. Note that spectral peaks occur at approximately 13 kHz at 23 days (prehearing), 28 days (onset of hearing), and 32 days (posthearing); however, calls at 32 days are very infrequent. Calls were recorded with a Bruel and Kjaer Type 4136$^{1}/_{4}$″ microphone and analyzed with a Stanford Research Systems Spectrum Analyzer; upper frequency limit for linearity approximately 80 kHz. (L. Aitkin, S. Cochran, and S. Frost, unpubl. observ.)

Pouch-young opossums do not utter adult screeches or growls prior to 150 days (McManus 1970). Instead the young utter a screech sound that has an extremely wide range of frequencies, with major components at 4, 12, 18, and 26 kHz. This type of vocalization is no longer given after the young has left the mother. This range of frequencies spans the most sensitive part of the audiogram (Fig. 3.1) and so should be very audible to the mother.

An important function of the auditory system of these marsupials may thus be the localization of helpless young which have strayed from the mother. It might

39

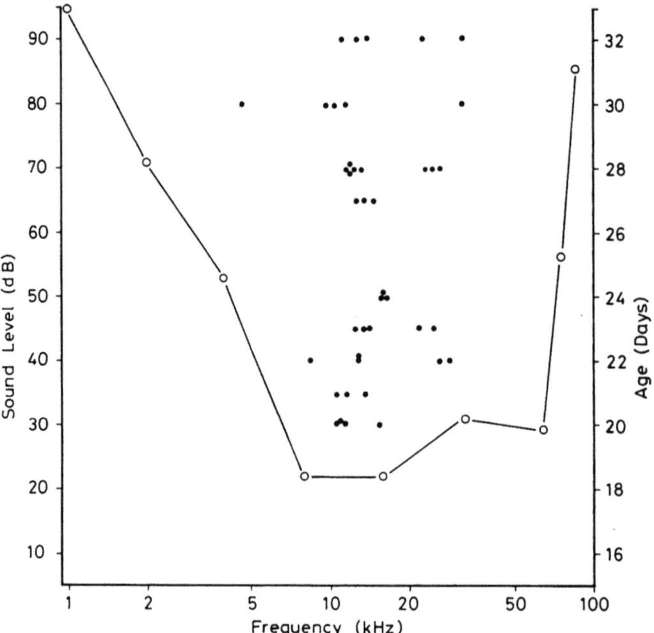

Fig. 4.7. Peaks of power spectra of the vocalizations of neonatal *Monodelphis* opossums (*filled circles*) plotted in conjunction with the behavioral audiogram of this species from the study of Frost and Masterton (1994; data kindly supplied by the late Dr. R.B. Masterton). Note the tendency for vocalization peaks to occur near the best frequency of adult hearing (10–20 kHz)

be expected that the hearing range of the mother is particularly sensitive to the sounds of the young and vice versa. It is well known from studies of the hearing of avians that hatchlings are more responsive to the maternal calls of their own species than to those of related species (Gottlieb 1971). Electrophysiological data have shown a corresponding sensitivity in the evoked potential audiograms of neonatal chicks and ducklings, with the peak auditory sensitivity in the neonate correlating with the peak energy output of the maternal call (Saunders et al. 1974).

4.5 Hearing Sensitivity and the Sounds of Prey and Predators

Apart from predation by humans, dingos, and (for smaller species) pythons and perhaps foxes (the latter introduced to Australia after white settlement), adult macropodids, particularly kangaroos, are sufficiently large and fleet of foot to be relatively immune from natural predators. Since the extinction of the Eastern quoll on the Australian mainland, arboreal marsupials also have few natural predators. Snakes and owls, particularly the powerful owl, are known to prey upon possums.

There is a much greater variety of predators for the smaller species, particularly those in South America, and in Australia the dasyurids, such as the Northern quoll, may be added to the list, along with another introduced species, the (feral) cat. The latter is the most intensively studied terrestrial predator in terms of hearing and hears over an enormous range of 10 octaves, from about 50 Hz to perhaps 90 kHz (Heffner and Heffner 1985a). The vocal repertoires of cats are only moderate compared with, for example, songbirds or monkeys, and much of the energy of their vocalizations lies well below their best frequency of hearing (8 kHz), with some of the higher harmonics of kitten calls reaching up towards the best frequency (Aitkin et al. 1994b). More than three octaves remain above the best frequency of cats' hearing that are not used in intraspecific communication.

Maximizing sensitivity to sounds of high frequency optimizes the detection of a particularly important environmental cue for sound localization, the interaural intensity difference. This arises because the head and pinnas act as an obstacle to sound, causing a shadowing effect on sound reaching the ear further from a sound source. The extent of head shadowing increases with both the frequency of the sound and the diameter of the head (Rayleigh 1876), so that animals with high-frequency hearing will have more of this cue to work with, if all other factors are equal. The extension of the cat's audiogram to high frequencies maximizes that species' ability to localize the sounds made by prey – the fluttering of wings, rustle of grass and twigs, and the high-pitched vocalizations of rodents, for example. These all contain a rich repertoire of high frequencies, especially transient sounds.

The hearing ranges of studied nocturnal marsupial carnivores appear not to extend to frequencies as high as that of the domestic cat, nor do those of dogs (Heffner 1983) or the smallest carnivore, the least weasel (Heffner and Heffner 1987). The heightened high-frequency sensitivity of cats may explain the great success of these terrestrial predators in all zoogeographic zones, whether they are native or introduced. Most dogs are diurnal and are likely to use vision (and olfaction) for prey detection rather than hearing.

The diet of small dasyurids such as the Northern quoll includes insects, lizards, frogs, small mammals (such as rodents), and carrion (Morton et al. 1989). Prey move through dry grass in the dry season, making crackling sounds. In the wet season the vocalizations of frogs and insects fall well within the range of the quoll's hearing. The pinna of the Northern quoll is highly directional for sounds above 15 kHz, so that additional cues are available for frequencies at the upper end of the quoll's hearing range (Aitkin et al. 1994b). Predators are dingoes and snakes; owls can also take younger quolls. It is not clear how sensitive quolls are to the sounds made by these predators.

Finally, for any species, the sounds made by conspecifics and by prey and predators will be detected against a background of abiotic noise generated in the environment in which that species lives. Thus the sound of heavy rain, of strong winds, of thunder, and of water flowing in a cascade could mask the biological sounds important to the species' survival, as could the acoustic density of the habitat (e.g., forest vs desert). There is some evidence, however, that features of vocalizations have evolved to meet the conflicting abiotic environmental influences (e.g., Brown and Waser 1988). In this regard, the acoustic spectrum of the nocturnal environment in which the Northern quoll hunts has very little energy at the best frequency of hearing of the quoll, so that the isolation calls of the young are unlikely to be greatly masked by the sounds of the environment (Aitkin et al. 1994b).

4.6 Future Research Directions

There has been considerable onomatopoeic classification, particularly in relation to social and behavioral circumstances, of the calls made by animals, but rather less work has been done on the spectra of vocalizations and very little on marsupials. From our knowledge of hearing ranges we can guess that some sounds generated by an animal may not be audible to conspecifics, whereas others will have most energy in the most sensitive part of their hearing range. It also seems clear that the success of a species will be facilitated by the sensitive detection of sounds made by prey and predator. Much more work needs to be carried out in the field to correlate hearing with these biologically important sounds generated in the normal abiotic acoustic background in which the species lives.

It is particularly important to have correlative information on both hearing and vocalization, since this will help develop general hypotheses about the "design" of the auditory systems of all species. For example, the tammar wallaby is being used more and more as a neurodevelopmental model for sensory systems, including hearing (Liu et al. 1996), but little is known of the auditory and vocal behavior of even this well-studied species.

Chapter 5

Auditory Periphery of Marsupials

5.1 The Outer Ear

Most mammals have prominent external ears (pinnae), moles and monotremes being exceptions. The conical shapes of many are clearly suited to the directional amplification of sound, although they also have other important functions, such as heat exchange and deflection of irritating insects. Variations in the shape of the pinnae of marsupials can be seen in Fig. 5.1. Some are pointed (e.g., tammar wallaby, Fig. 5.1A), others are more rounded (e.g., sugar glider, Fig. 5.1B). The inner folds and grooves may be quite simple (e.g., yellow-bellied glider, Fig. 5.1C) or complex, as in the case of dasyurids (e.g., kowari, Fig. 5.1D) and *Monodelphis* (Fig. 5.1E), in which the tragus (arrows, Fig. 5.1E) is bat-like.

Acoustical properties of the pinna have been examined in tammar wallaby and quoll cadavers by inserting probe microphones into the ear canal from within the head to the level of the eardrum and moving a loudspeaker around the head to different elevations (above and below the horizontal plane of the animal's head) and azimuthal positions (to the left and right of the midsagittal plane).

The pinnae of the quoll are fairly immobile, and changes in head position serve to reorient the direction in which they point (Aitkin et al. 1994b). The acoustical axis of the quoll pinna has, for frequencies above 15 kHz, a relatively invariant *azimuthal* position at 25–30° from the midsagittal plane and an elevation that is frequency-dependent. The maximum gain of the quoll's pinna is 25 dB near 40 kHz (Aitkin et al. 1994b). Prey-searching behavior in quolls involves the animal rapidly moving its head in a horizontal plane, parallel to the ground, with the nose and chin whiskers virtually touching the ground. The pinnae slope back relative to the lowered head position and so are in a position to scan the acoustic environment at the level of the horizon.

The tammar wallaby's pinna, which is very mobile in life (Coles and Hill 1981), has an acoustic axis constant in *elevation*, close to the horizontal plane for natural ear positions, with the azimuth determined by the pinna orientation on the head. For a fixed pinna position there is only a weak relationship between the spatial location of the acoustical axis and sound frequency, but the amplification depends on frequency, and the pinna has a maximum on-axis gain of 25–30 dB near 5 kHz (Coles and Guppy 1986). The direction-dependent gain of the pinna influences the important sound localization cue, the interaural intensity difference, which is mainly determined by head size and sound frequency. The computation of this cue is even more complex when the pinna is mobile, as it is for example with domestic cats and tammar wallabies.

In addition to the direction-dependent amplification of the external ear there is a frequency-dependent amplification due to the resonant properties of the ear

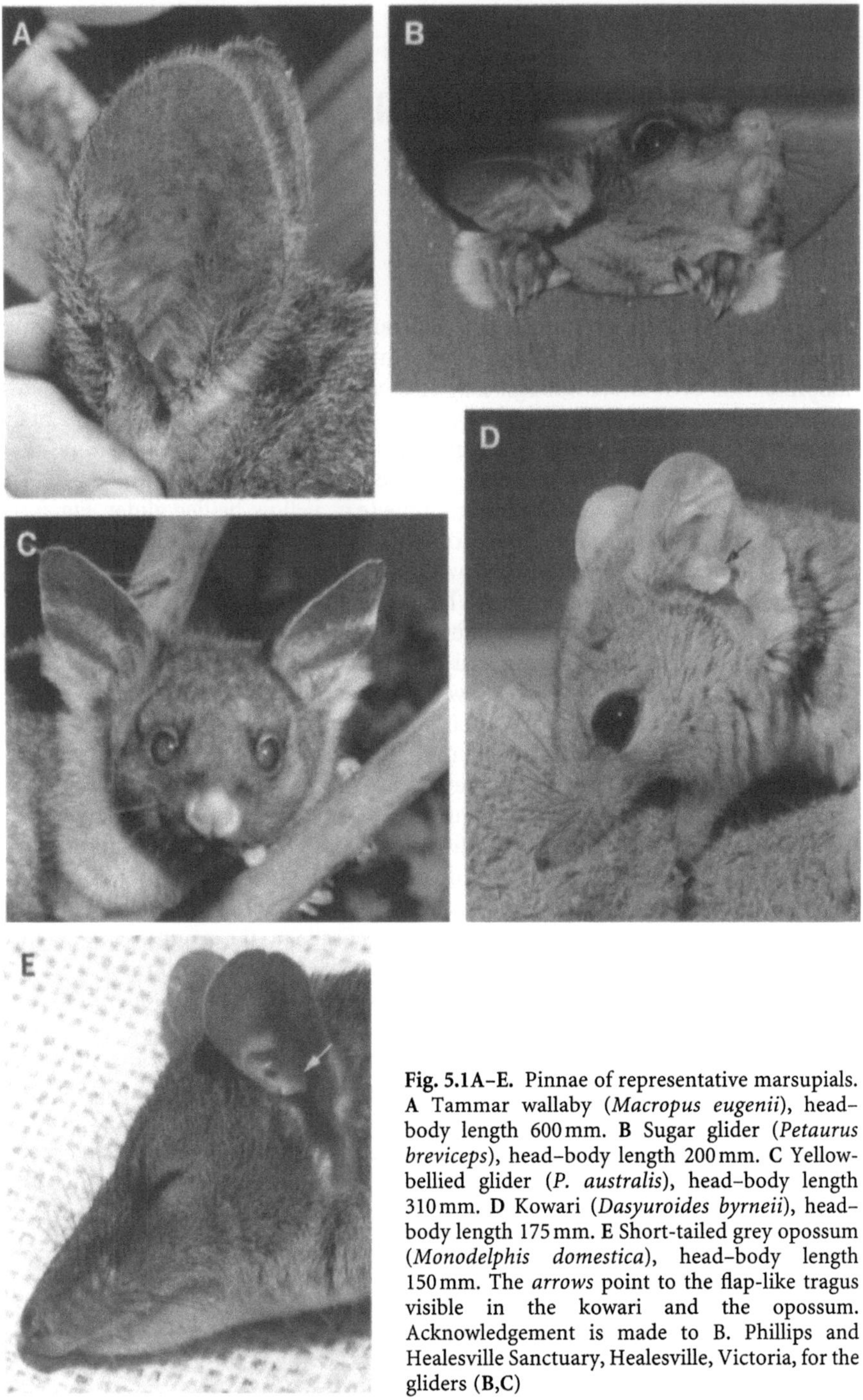

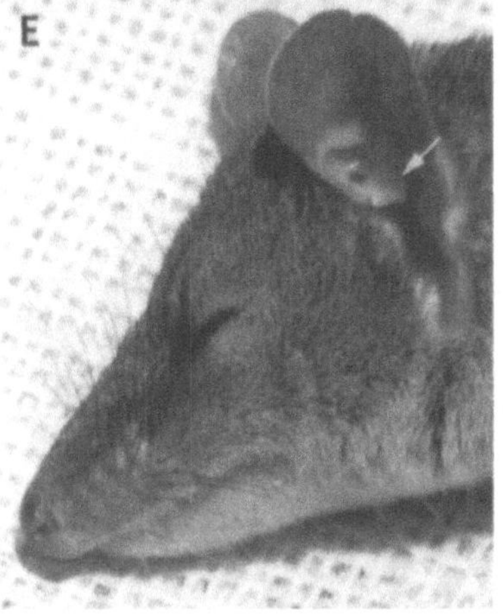

Fig. 5.1A–E. Pinnae of representative marsupials. **A** Tammar wallaby (*Macropus eugenii*), head–body length 600 mm. **B** Sugar glider (*Petaurus breviceps*), head–body length 200 mm. **C** Yellow-bellied glider (*P. australis*), head–body length 310 mm. **D** Kowari (*Dasyuroides byrneii*), head–body length 175 mm. **E** Short-tailed grey opossum (*Monodelphis domestica*), head–body length 150 mm. The *arrows* point to the flap-like tragus visible in the kowari and the opossum. Acknowledgement is made to B. Phillips and Healesville Sanctuary, Healesville, Victoria, for the gliders (**B,C**)

canal. For a tammar wallaby with an ear canal length of 1.6 cm (Coles and Guppy 1986), the amplification would be greatest near 1000 Hz; for the quoll, which is much smaller in size, the resonant frequency would be correspondingly higher. The ear canal of the feathertail glider is partly occluded by a disc of bone (Segall 1971; Aitkin and Nelson 1989), which has the effect (as illustrated in Fig. 4.3) of reducing the overall sound input to the glider's eardrum, with a relative enhancement near 25–30 kHz (Aitkin and Nelson 1989).

The tragus of dasyurids and certain other nocturnal marsupials expands to form a baffle external to the opening of the external ear canal (Fig. 5.1D,E). This feature is found in many echo-locating bats, particularly those which have more or less immobile pinnae (Pye 1968). Experiments on *Eptesicus* have shown that interference with the prominent tragus of this bat increases the angle of vertical perception from about 3° to 12–14° (Lawrence and Simmons 1982); these authors argue that the pinna and tragus provide a cavity for a strong secondary echo which, in conjunction with the primary signal, apparently encodes the vertical direction of a sound. It is possible in quolls that the enlarged tragus, which is largely horizontally disposed (Aitkin et al. 1994b), contributes to the variation in elevation, rather than azimuth, of the acoustical axis of the pinna at higher frequencies. It could be speculated further that nocturnal hunters with mobile ears have no need for a specialized structure to impart vertical directionality to the ear, since they can move their ears to home in on the sound without taking their eyes from a visual target. In support of this, the pinna of the striped possum is capable of rapid, fluttering movements and lacks an enlarged tragus (L. Aitkin, unpubl. observ.). On the other hand a very prominent tragus is present in *Monodelphis* (Fig. 5.1E), which displays very similar rapid flicking movements (L. Aitkin, unpubl. observ.).

5.2 Middle Ear Structures

5.2.1 Evolution of Middle Ear Bones

The mammalian middle ear is composed of three delicate interconnected bones – malleus, incus, and stapes – connecting two membranes, the tympanic membrane and the oval window of the cochlea (Fig. 2.1). The three-bone system is present in all three mammalian orders, but the fused malleus and incus of monotremes is heavy (Fig. 5.2A) and firmly attached to the periotic bone, making the middle ear system stiff at all frequencies compared with eutherians and at low frequencies compared with modern reptiles (Aitkin and Johnstone 1972; Gates et al. 1974).

Reptilian and avian ears have a single ossicle, the columella. This single column of bone carrying vibrations from the eardrum to the cochlea is the precursor of the stapes, which in most mammals is stirrup-shaped. The columella-like stapes of monotremes has been considered a primitive trait by some authors, and the physiological evidence in monotremes is suggestive of peripheral auditory systems that are less efficient than those of either eutherians or birds (Aitkin and Johnstone 1972; Gates et al. 1974).

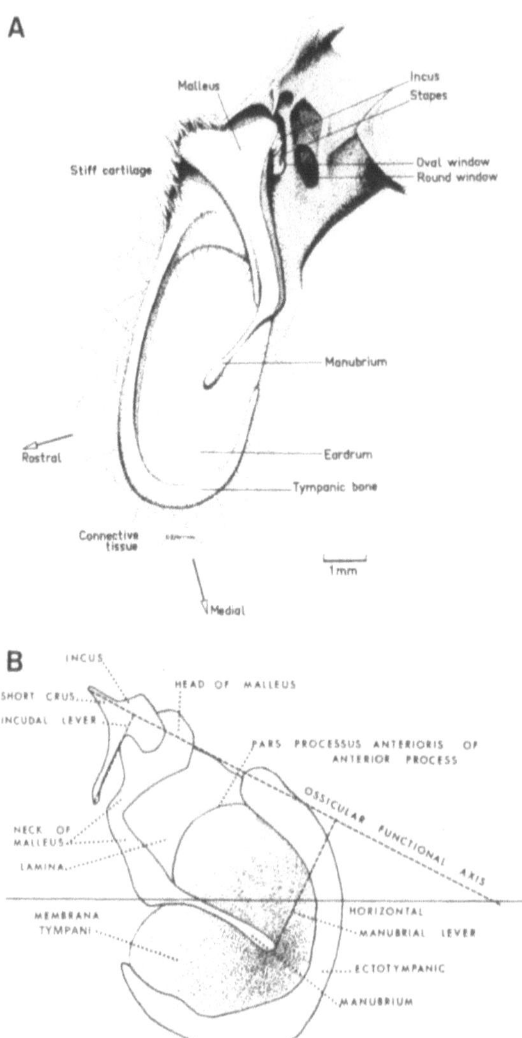

Fig. 5.2. **A** View of the middle ear of the platypus (*Ornithorhynchus anatinus*) from within. Note the tympanic bone suspended in connective tissue, the relatively massive malleus, and small incus and stapes (Gates et al. 1974). **B** Schematic illustration for marsupials of the ossicular functional axis, the malleolar, and incudal lever arms. (Segall 1969, with permission)

The malleus and incus are homologous with the articular and quadrate bones of the lower jaw of modern reptiles. At the stage in evolution when mammal-like reptiles appeared, the articular and quadrate bones were released from their original functions of supporting and forming the lower jaw and were incorporated into middle ear mechanisms. How and why this occurred is a matter of debate (see, e.g., Hopson 1966; Tumarkin 1968). The development of a three-ossicle system improved the amplification imparted by the middle ear since the three bones, acting as a lever, could contribute extra gain to that provided by the area difference between eardrum and stapes footplate. The presence of three bones rather than one does not fit easily the fact that mammals (excepting monotremes) have a much higher frequency response than birds (excepting owls) or reptiles, since the extra bones of mammals require extra mass and coupling.

5.2.2 Middle Ear of Marsupials

The fact that the middle ears of some marsupials are not housed in a bulla but are separated from other structures laterally and caudally by soft tissues has been mentioned previously (Chap. 3). Bullae vary in size and construction among eutherians. Larger bullae will act particularly as low-frequency resonating chambers, so that animals with low-frequency sensitivity (e.g., gerbils, kangaroo rats) are likely to have large bullae and vice versa. Following this line of reasoning, the lack of a bulla in marsupials would suggest that they are unlikely to have marked low-frequency sensitivity.

Segall (1969) surveyed the auditory ossicles of marsupials. He described, as had Doran (1877), the presence of three ossicles and a tympanic bone, to which the eardrum is attached (Fig. 5.2B). The latter is a separate bone in *Didelphis* and *Caenolestes*, but is incorporated into the ear canal in many other marsupials.

The ossicular functional axis is an imaginary line about which the middle ear bone system rotates; in Fig. 5.2B, up-and-down movements of the eardrum (membrana tympani) are coupled to the manubrium and neck of the malleus. The fulcrum for the lever action is the malleo-incudal joint, so that the movement of the malleus raises and lowers the incudal lever, which in turn moves the stapes (not shown in Fig. 5.2B) in and out of the oval window, the latter lying in the plane of the figure. The two levers are shown as dashed lines perpendicular to the ossicular functional axis, and the lever ratio of the system can be calculated from these distances.

Segall showed that both the angle of the ossicular functional axis to the horizontal plane and the lever ratio varied across the marsupial population studied. The functional significance of the former is not clear, but an increased lever ratio could lead to improved impedance matching and thus better hearing, all other factors being equal. The ossicular axis varied from $+23°$ in *Didelphis* to $+13°$ in *Dasyurus* to about zero in *Macropus* to $-5°$ in *Dromiciops*. Segall argued that this was a progression from generalized to specialized species and that this was also true for lever ratios, the smallest (1.4) being observed in *Didelphis* and the largest (2.2) in *Dromiciops*. The significance of these observations requires knowledge of the overall impedance matching of marsupial ears. For comparison, the lever ratio of the human ear is about 1.3 (von Békésy 1960) and that of the cat more than 2.2 (Tonndorf and Khanna 1967).

As in eutherians, two middle ear muscles have been observed in *Monodelphis* – the tensor tympani attached to the malleus and the stapedius attached to the stapes (Filan 1991). In eutherians the stapedius muscle is innervated by the facial nerve (7th cranial nerve) and the tensor tympani by the trigeminal nerve (5th cranial nerve). Contraction of either muscle increases the stiffness of the ossicular chain. They contract in response to loud sounds (less than 1.5 kHz, greater than 80 dB) and as part of coordinated neck muscle contractions (chewing, vocalizing, swallowing; see, e.g., Dallos 1984). Nothing has been published about their functional properties or innervation in marsupials.

5.3 Cochlea

5.3.1 Evolution of Sound Reception

In all mammalian ears, the basic receptor elements, hair cells, are supported on a movable platform, the basilar membrane, the whole being immersed in a fluid-filled vesicle, the cochlea. The stereocilia (hairs) of the hair cells are bent due to the differential motion of basilar membrane and fluid, and the ensuing depolarization of the hair cell causes the release of a chemical transmitter which depolarizes the attached cochlear nerve afferents. A number of steps have been taken during evolution to reach this goal (Wever 1974).

In fish, macular organs developed in which otoliths moved sensory cells. These organs signal both gravity and sound vibrations to separate nerve supplies. Two organs, the amphibian and basilar papillae, appeared in amphibians. These signal sound and gravity separately although both depend on membranes acting on hair cells for transduction. In reptiles, for the first time, hair cells rested on a structure moved by fluid. Birds and mammals evolved from reptiles along different paths, the latter stemming from mammal-like (therapsid) reptiles.

In mammals other than monotremes, the cochlea is an elongated, coiled structure, the number of coils varying from 1.5 in hedgehogs to 4 in guinea pigs. The cochlea of monotremes is a wide, sickle-like curve which accommodates three rows of inner and six rows of outer hair cells, and there is an additional chamber at the distal end of the cochlear duct which is similar in appearance to the lagena of birds (Smith and Takasaka 1971). The total number of outer hair cells is much smaller than in eutherians, but the numbers of inner hair cells are comparable in the two orders (Ladham and Pickles 1996). Fernández and Schmidt (1963) discussed the evolution of the coiled cochlea, noting the lack of evidence for coiling in the earlier therapsid reptiles and in modern monotremes, and consider it to have appeared sometime between late Triassic and late Jurassic (170–140 million years B.P.). Recent work by Meng and Fox (1995a,b) places the first appearance of a coiled cochlea in the Late Cretaceous, being present at about the same time in marsupial and placental fossil material.

5.3.2 The Cochlea of Marsupials

Light microscope descriptions of the gross morphology of the cochlea have been published for *Didelphis* (McCrady 1938; Fernández and Schmidt 1963), *Monodelphis* (Willard and Munger 1988), both members of the order Didelphidae, and *Trichosurus* (Aitkin et al. 1979), from the Phalangeridae. Cochleas of four representatives of other orders, quoll (Dasyuridae), pademelon (Macropodidae), ringtail possum (Petauridae), and feathertail glider (Burramyidae), are illustrated in Fig. 5.3. The cochlea is completely contained within the temporal bone in the pademelon, ringtail possum, and quoll, but protrudes into the middle-ear cavity in the feathertail glider. The cochlea of *Monodelphis* has 1.75 turns, those of *Didelphis*, pademelon, feathertail glider, ringtail, and quoll have 2.5 turns, and that of the brushtail possum has 3.5 turns.

Three chambers, the scala vestibuli (SV), scala media (SM), and scala tympani (ST), are separated by the Reissners membrane (between SV and SM) and the basilar membrane (between SM and ST; Fig. 5.3). The cross-sectional area of each scala is greatest at the base, but the width of the basilar membrane increases from base to apex. The organ of Corti (OC) appears in these low-power cross sections as a hump on the basilar membrane (BM) where it meets the bony spiral (osseous spiral lamina). During preparation of surface views and for scanning electron microscopy (e.g., Fig. 5.5), the temporal bone is chipped away, along with external vascular membranes, to expose the bony spiral and organ of Corti. The distal processes of the cochlear nerve run from the organ of Corti to their cell bodies in the spiral ganglion (SG; visible in all sections but most prominent in the higher

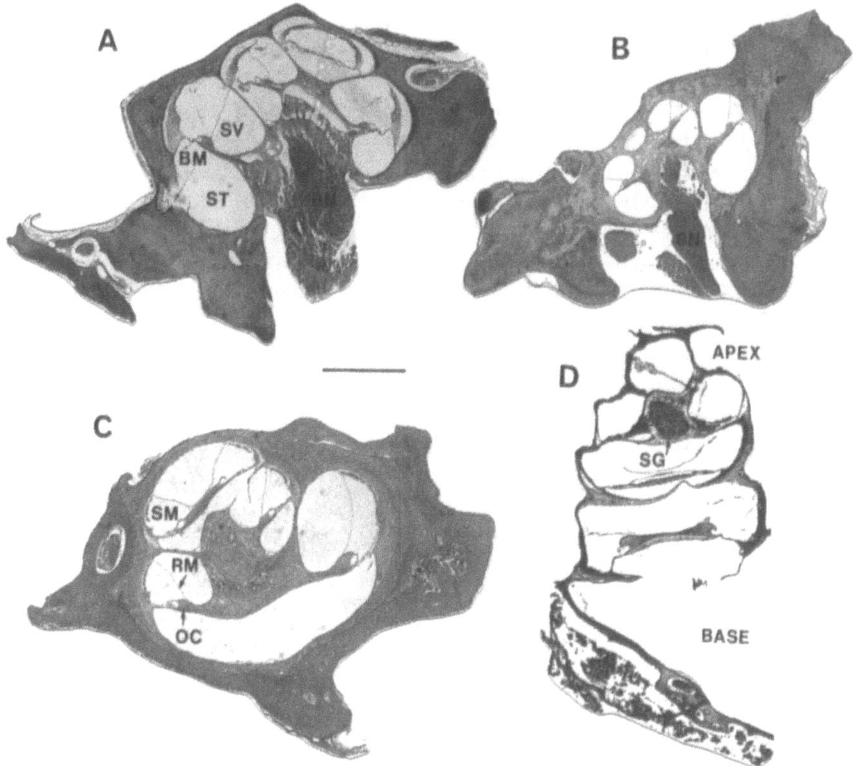

Fig. 5.3A–D. Photomicrographs of sections through the temporal bones of four marsupials: A Northern quoll; B pademelon wallaby; C ringtail possum; D feathertail glider. *SV, ST, SM* Scalae vestibuli, tympani, media; *BM, RM* basilar membrane, Reissners membrane; *SG* spiral ganglion; *OC* organ of Corti; *CN* cochlear nerve. Calibration: 1 mm (**A,C**); 2 mm (**B**); 0.56 mm (**D**)

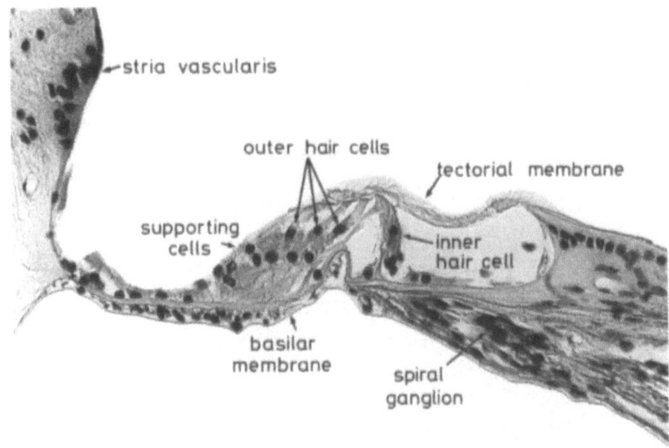

Fig. 5.4. Transverse section through the cochlea of the Northern quoll stained with hemotoxylin and eosin, revealing details of the organ of Corti

power view of the feathertail glider); proximal processes leave the spiral ganglion and join the cochlear branch of the eighth cranial nerve, which passes down the center (modiolus) of the cochlea.

Fernández and Schmidt (1963) consider both marsupials and placental mammals to exhibit the same type and distribution of cells in the organ of Corti. Thus, in the quoll, there are three rows of OHC and one of IHC, and pillar cells and Deiters cells are clearly recognizable (Fig. 5.4). The tectorial membrane is subtended from the bony limbus and overhangs the hair cells. The basilar membrane in Fig. 5.4 has an upward fold due to experimental artifact; its length from base to apex is approximately 14 mm. The length of the basilar membrane in the opossum is 15 mm and that in the brushtail possum 17 mm (Aitkin et al. 1979); these are comparable to that of the guinea pig (18 mm), but are somewhat shorter than that of the cat (22 mm). Cochlear length is partly related to the size of the animal (von Békésy 1960) – compare the above basilar membrane measurements in small animals to those of humans (35 mm), cows (38 mm), and elephants (60 mm). The relationship between location on the basilar membrane and frequency in *Monodelphis* has been examined by Müller et al. (1993), who found a frequency map very similar to those described for eutherians by von Békésy (1960).

Views, using scanning electron microscopy, looking down on the reticular lamina with the tectorial membrane removed, yield a very conventional cochlear structure in *Monodelphis*: at the base (Fig. 5.5A) are found regular V-shaped arrays of the stereocilia belonging to three rows of OHC and a single row of horizontally disposed IHC stereocilia. The stereocilia of the OHC are embedded in the tectorial membrane (Fig. 5.5B), whereas those of the IHC appear not to be. There is a rather more irregular arrangement at the apex of the cochlea as in eutherians and a complete lack of orientation in the stereocilia (Fig. 5.5C). Features such as efferent synapses on basal OHCs, lateral cysternae, and cysternal pillars under OHC membranes (correlated with motility; Pujol et al. 1997) and type I and II ganglion cells are present in *Monodelphis*, as they are in eutherians

50

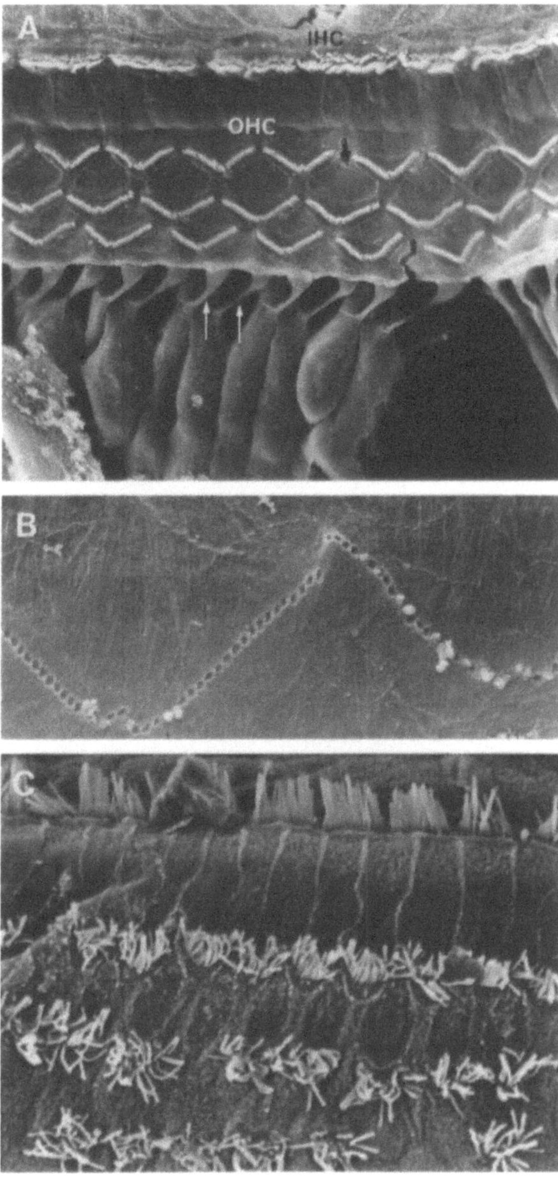

Fig. 5.5A–C. Scanning electron microscope views of surface preparations of the organ of Corti in adult *Monodelphis* opossums. **A** The regular arrays of the stereocilia of outer hair cells (*OHC*) may be contrasted with the more linear arrays of inner hair cell (*IHC*) stereocilia at the base of the cochlea. Note the supporting arms of Deiters cells (*arrows*); in *Monodelphis* each cell has two such arms. **B** The underside of the tectorial membrane from the base of the cochlea reveals the pit-like indentations in which OHC stereocilia were embedded. A few OHC stumps remain. No equivalent series of pits corresponds to IHC stereocilia. **C** The arrays of stereocilia at the apex of the cochlea are much less regular, particularly those from the OHC (*lower rows*). Original magnification of **A**, 900×; of **B**, 6000×, of **C** 1100× (L. Aitkin, R. Pujol, and G. Humbert, unpubl. observ.)

(unpubl. observ.). One variation on the eutherian scheme is the presence of Deiters cells, supporting the outermost OHCs, which have twin supporting pillars (arrows, Fig. 5.5A).

5.3.3 Cochlear Potentials in Marsupials

Reference has already been made in Chapter 2 to the methods of recording cochlear potentials. Fernández and Schmidt (1963) made round-window and intracochlear recordings in *Didelphis virginiana* and observed very similar electrical events to those measured in eutherians. A microelectrode in scala tympani recorded zero potential difference relative to the neck muscles. As the tip of the electrode penetrated the organ of Corti, presumably entering inner hair cells, the potential became about 35 mV negative, and continued penetration resulted in a potential reversal to more than 90 mV positive – the endocochlear potential observed in many placental species.

The amplitude of the cochlear microphonic potential (CM) was logarithmically related to sound pressure level, and peak amplitudes of about 500 μV were recorded in opossums, compared with 1.5–2.0 mV for cats and monkeys. Similar results have been obtained with the CM potentials of brushtail possums (Aitkin et al. 1979). It may be that, for any given sound-pressure level, CM potentials are smaller in marsupials than eutherians, but the difference may also be due to technical considerations related to the lack of an auditory bulla in marsupials and the distance of the recording electrode from the active sites in the cochlea.

Summating potentials and compound cochlear nerve action potentials have also been recorded in *Didelphis* (Fernández and Schmidt 1963). The former appears as the baseline voltage upon which the CM sits, and, like the CM, it is smaller than summating potentials recorded in eutherians.

5.4 Future Research Directions

The peripheral auditory structures of marsupials belong to the spectrum of arrangements found in eutherians; nothing specifically "marsupial" is apparent. More, however, could be learned about the middle-ear amplifier and middle ear muscles of marsupials. The fact that the three marsupials studied behaviorally (*Didelphis, Monodelphis, Marmosa*) have higher thresholds than eutherians prompts behavioral investigation of members of the Diprotodontidae, and examination of the effectiveness of the middle ear bones and the tympanic bone in sound transmission in the Didelphidae. The ultrastructure of the cochlea and its afferent and efferent connections have only just begun to be examined. Could the OHC system in the Didelphidae also contribute to their poorer thresholds? In particular, it would be of great interest to characterize the synapses on the cochlear hair cells and the sources of efferents to these cells, and to extend the observations on cochlear potentials to more marsupials using more modern techniques.

Chapter 6

Auditory Structures of the Brainstem

6.1 The Marsupial Brain

Marsupial brains are characterized by a generally smooth neocortex with few fissures (lissencephaly) and a prominent gap between the cerebrum and cerebellum, through which the superior and inferior colliculi are visible (Fig. 6.1A). Lissencephaly is observed in many eutherian species, from insectivores to primates, and so cannot be considered a "primitive" trait. The size of the brain relative to the body gives some insight into the development of the brain, or "encephalization", of a species. Larger animals generally have larger brains, although small species have relatively larger brains than larger species. Stephan (1972) used an allometric equation that accommodated for the nonlinearity of the brain weight/body weight relationship, and Nelson and Stephan (1982) applied this to marsupials, mostly Australian species.

The extent of the encephalization of marsupials overlaps with insectivores, but their indices are higher than those of basal insectivores, and a few species overlap with the lowest end of the prosimian range. Within Marsupialia some species stand out as progressive, including the petaurids *Dactylopsila* (striped possum) and *Gymnobelideus* (leadbeater possum); in contrast, the species of *Planigale* have indices well below those of the parent family, Dasyuridae. Nelson and Stephan argue that encephalization indices would place marsupials above the grade of basal insectivores, at about the grade of progressive insectivores and lower prosimians, in evolution. In this sense, the brains of marsupials are not primitive.

An anatomical description of the entire brain of *Didelphis* (mostly the southern American opossum *D. marsupialis*, but some *D. virginiana* material was available) has been published in conjunction with stereotaxic coordinates (Oswaldo-Cruz and Rocha-Miranda 1968). The gross subdivision of the auditory pathway in marsupials is similar to eutherians (see Chap. 2 and Fig. 2.3). A view of the dorsal surface of the brain of a Northern quoll (Fig. 6.1A) reveals a cerebral cortex largely devoid of significant fissures, with shallow grooves, in which lie pial vessels, separating functionally distinct regions of cortex. The dorsal structures of the tectum – the superior and inferior colliculi – are located in a noticeable space between the cerebrum and cerebellum (Fig. 6.1A,B). When the cerebellum is removed on one side (Fig. 6.1B; brushtail possum) the cochlear nucleus (CN) and eighth cranial nerve (8) can be seen surrounded by the stalks of the cerebellum peduncles (RB, BC, BP). More rostrally the medial and lateral geniculate bodies are visible (MGB, LGB) caudal and ventral to the auditory cortex (AC).

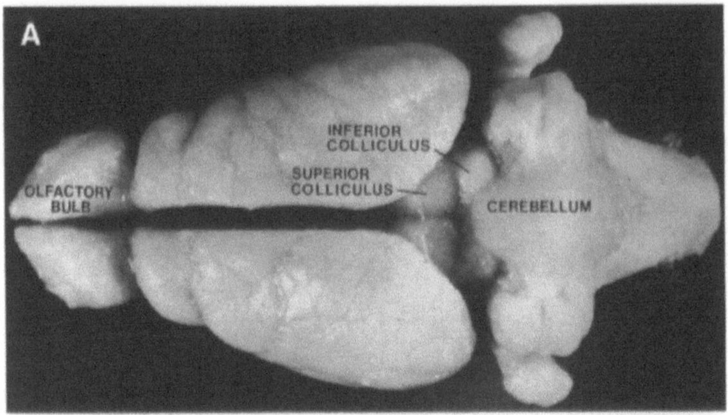
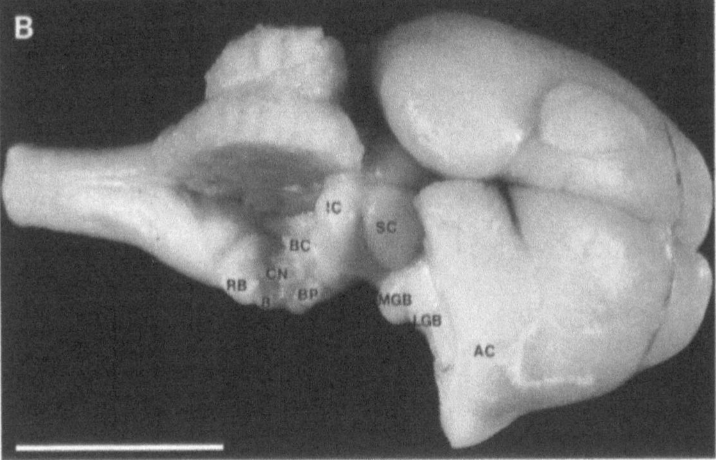

Fig. 6.1. **A** Brain of Northern quoll showing lissencephalic cerebral cortex, large olfactory bulbs, exposed superior and inferior colliculi, and small cerebellum. **B** Brain of brushtail possum with occipital cortex and cerebellum removed on right side to expose brain stem and thalamic structures. *8* Eighth cranial nerve; *CN* cochlear nucleus; *RB, BP, BC* three cerebellar peduncles: restiform body, brachium pontis, brachium conjunctivum; *IC, SC* inferior and superior colliculi; *MGB, LGB* medial and lateral geniculate bodies; *AC* auditory cortex. Calibration: **A** 10 mm, **B** 15 mm

6.2 Auditory Nuclei of the Medulla

6.2.1 Cochlear Nuclear Complex

A major difference between the brain-stem auditory pathways of marsupials and eutherians, first observed by Stokes (1912), is in the disposition of the cochlear nuclei. In eutherians the cochlear nuclei are largely external to the restiform body (Fig. 6.2B), whereas in marsupials they are almost entirely internal (Fig. 6.2A). This general feature of the cochlear nucleus has been recognized in a number of more recent studies (red kangaroo, Cowley 1973; various, Johnson 1977; Virginia

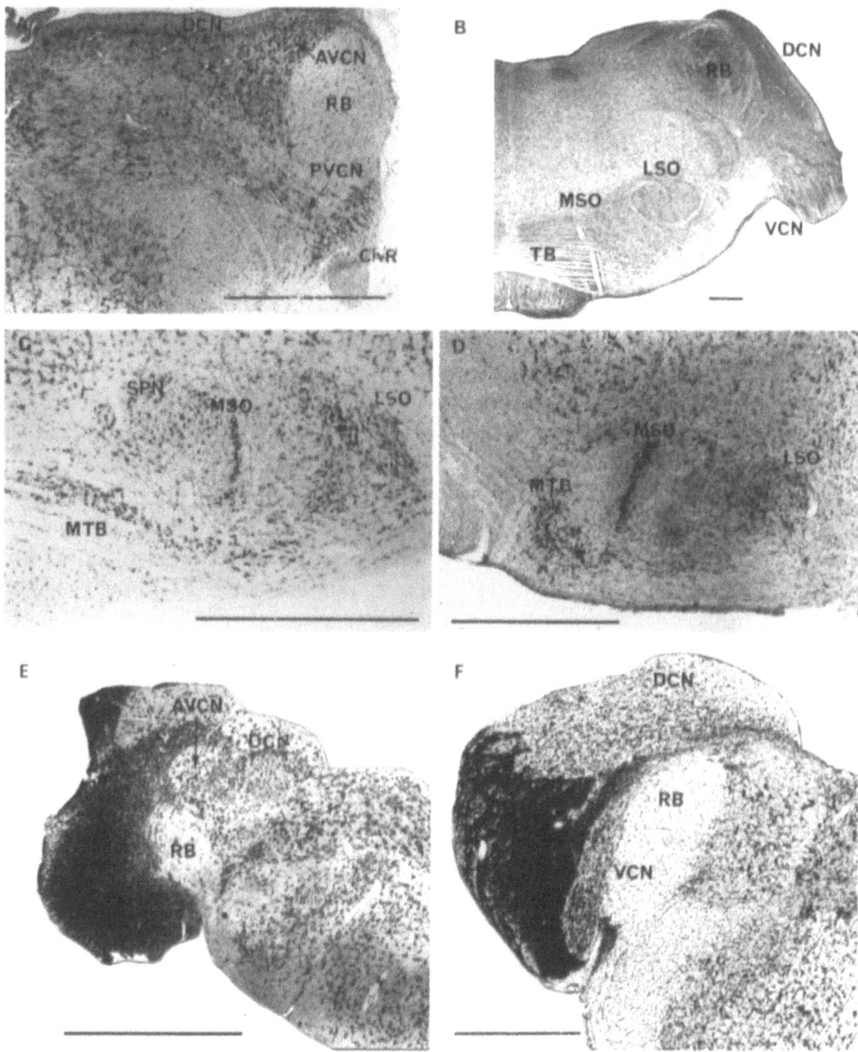

Fig. 6.2A–F. Photomicrographs of brain-stem regions in marsupials (*left*) and eutherians (*right*). **A** Cochlear nuclear complex in the Northern quoll. **B** Medullary nuclei in the domestic cat. **C,D** Superior olivary complex in the ring-tailed possum and common marmoset monkey, respectively. **D,E** Hypertrophied cochlear nuclear complex in the feathertail glider and mountain beaver, respectively. *DCN, AVCN, PVCN* Dorsal, anteroventral, and posteroventral cochlear nuclei (in **B** and **F** these subdivisions are not distinguished but are collected together as the ventral cochlear nucleus, *VCN*); *CNR* cochlear nerve root nucleus; *LSO, MSO* lateral and medial superior olivary nuclei; *MTB* medial nucleus of trapezoid body (*TB* in **B**); *SPN* superior paraolivary nucleus. Note the position of the restiform body (*RB*) lateral to the cochlear nuclear complex in **A,E** and medial in **B,F**. Calibration bar for all frames is 1 mm. (Aitkin 1995, with kind permission of Elsevier Science – NL, Sara Burgerhartstraat 25, 1055 KV Amsterdam, The Netherlands)

opossum, Willard and Martin 1983; brushtail possum, Gates and Aitkin 1984; quoll, Aitkin et al. 1986a; nine species of possums and gliders, Aitkin 1996) and is a characteristic of all marsupial families. The factors responsible for these differing arrangements in eutherians and marsupials are not known but may relate to the very lateral extension of the cerebellar plate relative to the lateral medulla during development in marsupials (e.g., Aitkin et al. 1994a). In spite of this gross difference, recognizable to the naked eye, the subdivisions of the cochlear nuclei may all be recognized – the laminated dorsal nucleus (DCN, Fig. 6.2A,E), the tightly packed cells of the anteroventral nucleus (AVCN, Fig. 6.2), and the islands of cells that collectively form the posteroventral nucleus (PVCN). As is the case with rodents (e.g., Merchán et al. 1988), marsupials have a distinctive cochlear nerve-root nucleus (CNR, Fig. 6.2A).

The cochlear nuclei of possums and gliders have been studied in some detail (Aitkin 1996). With the exception of the honey possum, the DCN is always the largest of the cochlear nuclei and is as great or greater in volume than the sum of the two ventral cochlear nuclei in the brushtail possum, greater glider, and sugar glider; the AVCN extends throughout almost the entire rostrocaudal extent of the complex. The proportion of brain weight represented by the cochlear nuclear complex is greatest in the smallest four marsupials studied (honey possum, feathertail glider, sugar glider, and leadbeater possum) compared with larger species.

A very unusual cochlear nucleus has been observed in the feathertail glider (Aitkin and Nelson 1989). The dorsal cochlear nucleus expands into a structure called the lateral lobe which is richly supplied with granule cells (Fig. 6.2E). The expansion is lateral and caudal to the restiform body (RB). It is not clear whether this structure has auditory functions, although fibers of the cochlear nerve can be traced into the lateral lobe. This specialization is absent in other gliding marsupials (Aitkin 1996), but there is a striking morphological similarity between the lateral lobe of the feathertail glider and the hypertrophied dorsal cochlear nucleus of a burrowing rodent, the mountain beaver (Fig. 6.2F; Merzenich et al. 1973). In that species neurons in the enlarged magnocellular DCN have strong auditory response properties. The cochlear nucleus of another burrowing rodent, the pocket gopher, has a large development of cochlear granule cells but a normal dorsal cochlear nucleus (Merzenich 1970). This specialization is an unusual example of convergence between marsupials and eutherians.

Three fiber tracts arise from the eutherian cochlear nuclear complex – the dorsal, intermediate, and ventral acoustic striae (trapezoid body) – with targets of the IC, periolivary nuclei, and SOC/IC, respectively. There is only one study of a marsupial (the Virginia opossum) in which the efferents of the cochlear nucleus have been directly traced, either by lesion of the cochlear nucleus or injection of an anterograde tracer (Willard and Martin 1983). In this study, no evidence was found for an intermediate acoustic stria, although both dorsal and ventral striae were identified. Some fibers of the VCN join those of the DCN to form a large dorsal acoustic stria, which is also obvious in other marsupials (e.g., brushtail possum; Aitkin and Kenyon 1981).

6.2.2 Superior Olivary Complex

The nuclei of the superior olivary complex of eutherians can also be recognized in marsupials – lateral and medial superior olivary nuclei (Fig. 6.2B, LSO and MSO), ventral and medial nuclei of the trapezoid body (Fig. 6.2B, MTB), and, as in rodents (e.g., Harrison and Feldman 1970), a superior paraolivary nucleus composed of large multipolar cells (Fig. 6.2B, SPN; Aitkin and Kenyon 1981; Willard and Martin 1983; Aitkin et al. 1986a). In contrast to the cochlear nuclei, the individual nuclei of the superior olivary complex in marsupials have identical positional relationships to those of eutherians (e.g., marmoset, a primate, Fig. 6.2D). This observation has been confirmed by the present author for 12 closely studied marsupial species. In possums or gliders (nine species were studied in detail; Aitkin 1996) the LSO always appears as the largest nucleus of the superior olivary complex and, with the exception of the feathertail glider, is larger than the combined volumes of MSO and MTB. Morphometric analyses of neuronal populations carried out on the MSO and MTB show that the smaller species have a greater proportion of neurons relative to brain weight in both MSO and MTB than the larger species.

It is not clear what adjustment should be made in relating cell numbers to the size of the brain and body. Large animals are likely to have more cells in the MSO than smaller species, but as for brain vs body size (Nelson and Stephan 1982), it may not be a linear relation, and there may be a minimum number of cells a nucleus needs for it to operate functionally. If the latter has not been reached, it would appear that the MSO is disproportionately enlarged in the feathertail glider, and it is generally better developed in the four smaller marsupials (feathertail glider, honey possum, sugar glider, and leadbeater possum) than in the larger animals; that is, a greater proportion of the brain is devoted to the MSO in these smaller animals. It should be noted, however, that the MSO of most arboreal marsupials appears small compared with those of arboreal eutherians (Aitkin 1996).

Most workers in the field would accept that the superior olive has an important role in sound localization, but there are a number of hypotheses about the precise nature of this role. Some of the most suggestive data comes from comparative studies (Irving and Harrison 1967; Heffner and Masterton 1990). Irving and Harrison counted the numbers of neurons in the MSO, LSO, and MTB of 14 species. They also counted the number of cells in the abducens nucleus and measured the diameter of the eye. They found a strong correlation between the number of cells in the MSO and the two indices of vision, and suggested that the MSO may have a role in directing gaze towards the source of a sound. The MSO was largest, in their sample of species, in animals with large eyes, such as cats and primates. More recently, Heffner and Heffner (1992) have demonstrated a strong correlation between the azimuthal sound-localization threshold and the width of the field of best vision, reinforcing the idea that auditory and visual localization are strongly linked together.

The size of the MSO has also been correlated to the interaural distance and the extent of low-frequency hearing. Thus, animals with large interaural distances or well-developed low-frequency hearing also tend to have a large MSO (see e.g.,

Heffner and Masterton 1990), so that it is possible that the MSO has a role in the localization of low-frequency sounds. In relation to the relatively large size of the MSO in smaller arboreal marsupials, it may be that the smaller animals, being the potential prey of a larger field of predators (carnivorous marsupials, owls, snakes), need a greater ability (and thus more information) to detect a predator.

Irving and Harrison (1967) could find no correlation between the numbers of cells in LSO or MTB and visual indices, but noted that they were large in animals with good high-frequency hearing, such as bats, mice, and dolphins. There is a tacit assumption in many studies of sound localization that the important sounds to be localized are those of predator or prey. Less attention is given to the sounds generated by adult or neonatal conspecifics, i.e., complex vocalizations. These are clearly important for herbivorous arboreal mammals, such as certain primates and marsupials, and the fact the the LSO is usually the largest of the superior olivary nuclei in the arboreal marsupials may relate to their often highly vocal behavior and to the high pitched isolation calls generated by neonatal marsupials. For herbivorous arboreal marsupials, predator avoidance is still essential, but prey detection, hunting behavior, and eye/ear coordination may be less important.

Among marsupials, afferents to the superior olive from the cochlear nucleus have been studied only in opossums (Willard and Martin 1983). The MSO is innervated from the VCN of both sides, the LSO from the ipsilateral VCN and the MTB from the contralateral VCN. These connections are the same as those shown schematically for eutherians (Fig. 2.1). The SPN receives a heavy projection from the contralateral DCN (Willard and Martin 1983). Efferent projections have been studied in three species of marsupials using the retrograde transport of horseradish peroxidase from the inferior colliculus (IC). In the Virginia opossum, Willard and Martin (1983) report that the MSO projects equally to both IC, the LSO also projects bilaterally (but more strongly contralaterally), and the SPN supplies only the ipsilateral IC. This generally bilateral projection in the Virginia opossum of both LSO and MSO was confirmed in a double-labeling study (Willard and Martin 1984). In the brushtail possum the LSO projects mainly contralaterally, whereas the medial groups (equivalent to MSO and SPN) project mainly ipsilaterally (Aitkin and Kenyon 1981). In the quoll there is a clear tendency for lateral groups to project contralaterally and medial groups ipsilaterally (Aitkin et al. 1986a). These projection patterns differ in details from the eutherian pattern (Fig. 2.1), but more information is clearly needed about these and other brain-stem connections before meaningful comparisons can be made.

The bilateral projection of MSO to IC in opossums (Willard and Martin 1983) appears to differ from the largely ipsilateral eutherian pattern (Irvine 1986), and Kudo and his colleagues (1990) have suggested a phylogenetic trend in the laterality of this pathway. This may be true for the "primitive" Virginia opossum, but it does not apply to the two other marsupials studied, the brushtail possum and Northern quoll. Resolution may have to await investigations in other members of the Didelphidae, such as *Monodelphis*.

The superior olive is the source of the olivocochlear bundle, a pathway composed of efferent fibers destined for the inner and outer hair cells of the cochlea (see Chap. 2). These efferent pathways have been well described in eutherian laboratory animals (for review, see Rajan 1990; Warr 1992) and can be divided into a lateral system supplying the outer hair cells, originating from cells in and around the LSO, mostly on the ipsilateral side, and a medial system originating from cells in the medial part of the superior olive, mainly on the contralateral side.

The presence of efferent terminals in the cochlea of *Monodelphis* has been noted in Chapter 5, and the existence of the olivocochlear bundle in opossums has long been known. In his classic description of the superior olive, Rasmussen (1946) made lesions in the superior olive of opossums and concluded that the olivocochlear bundle in the opossum is very similar to that in the cat, except that it accompanies the outgoing limb of the facial motor root farther before crossing it, and rises up to pass over the restiform body before exiting the brain with the vestibular nerve. Clearly, more research needs to be done on the efferent control of the cochlea in marsupials.

6.3 The Auditory Midbrain

6.3.1 Cytoarchitecture of the Inferior Colliculus

The central nucleus of the inferior colliculus (CNIC) of eutherians is an ovoid mass which exhibits, when entire neurons with their dendrites are stained with the Golgi stain, a layered structure in which the dendrites of the principal bitufted cells are arranged along the trajectories of incoming afferent fibers (Rockel and Jones 1973a; Fitzpatrick 1975; Oliver and Morest 1984; Faye-Lund and Osen 1985). The CNIC is flanked laterally and rostrally by the lateral or external nucleus (EN), characterized by the presence of small neurons of variable dendritic orientation with a low packing density relative to CNIC, and dorsally and caudally by DC, which has broad loose layers of cells whose dendritic fields are more orthogonal to those in CNIC (Morest and Oliver 1984). In Nissl stains, cells in the IC of *Didelphis* are arranged in a compact ventromedial core (pars principalis) and a more loosely arranged, mainly laterodorsal shell (Oswaldo-Cruz and Rocha-Miranda 1968). Inspection of Nissl material in published studies suggests that the proportion of the inferior colliculus occupied by pars principalis (CNIC) is much smaller in marsupials than in cats.

More recently a tripartite division similar to that described above for cats has been described when the Golgi technique is employed in opossums (Willard and Martin 1983) and brushtail possums (Aitkin and Kenyon 1981). These studies and unpublished material from quolls show clear evidence of a laminar organization of CNIC (Fig. 6.3). In an adult brushtail possum the rows of neurons and glial cells change their orientation from lines parallel to the curved ventromedial surface of CNIC, through nearly straight lines from ventrolateral to dorsomedial, to curves parallel to the lateral surface of the IC (Fig. 6.3). The IC of the 60-day-old quoll (see Chap. 8), close to the onset of hearing in this species, shows that the laminar organization of the nucleus appears to be determined by incoming afferent fibers; the cells with their radiating dendrites show a tendency to line up along these penetrating fibers.

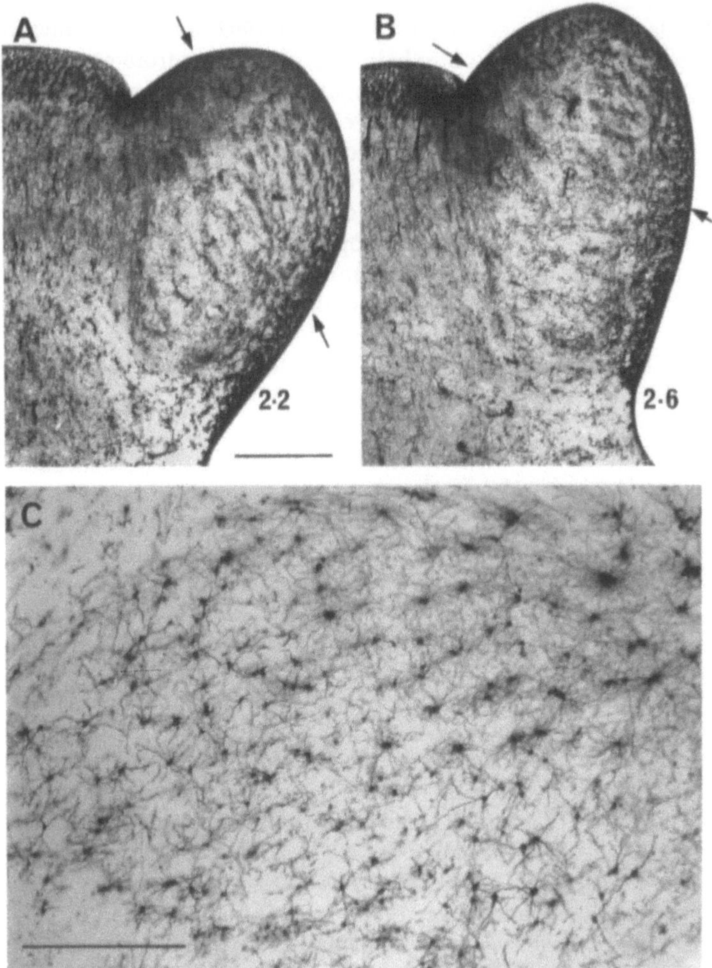

Fig. 6.3A–C. Sections impregnated with the Golgi technique showing the tendency of dendrites of neurons to be aligned in "laminae" in the inferior colliculus. **A,B** Sagittal sections from a brushtail possum cut at 2.2 mm (**A**) and 2.6 mm (**B**) from the midline. *Arrows* indicate approximate alignments of laminae of the inferior colliculus in each section. **C** Transverse section from 60 day Northern quoll showing at greater magnification some of the details of the rows of dendrites. Calibration: 2 mm (**A,B**), 500 μm (**C**)

6.3.2 Afferent Connections of the Inferior Colliculus

Auditory projections to the auditory midbrain of eutherians arise from many brainstem nuclei, from the opposite IC, and from the ipsilateral auditory cortex (Fig. 2.1; Roth et al. 1978; Adams 1979). Three major routes originate in the contralateral cochlear nuclear complex – that from AVCN via the SOC, from the AVCN direct, and from DCN. Another major source of afferents to the IC is the NLL.

Injections of retrograde tracer into the IC of marsupials label neurons with dendritic structures and nuclear locations that are similar to those in studies of eutherians (brushtail possum, Aitkin and Kenyon 1981; Virginia opossum, Willard and Martin 1983, 1986; quoll, Aitkin et al. 1986a). There are some differences between eutherians and marsupials, such as the "nuclear islands" in the dorsal acoustic stria projecting axons to the IC of brushtail possums (Aitkin and Kenyon 1981), the strong projections from the ipsilateral SPN, and the projections of CNR neurons to the auditory midbrain. The position of CNR in these species closely resembles the aggregation of cochlear root neurons (RN) described in rodents (for review, see Lopez et al. 1993). In rodent species, root neurons are particularly large, distinctive in shape, receive collaterals from cochlear nerve fibers prior to their main bifurcation, and project to the nuclei of the lateral lemniscus and superior colliculus.

The IC is a target for fibers descending from the auditory cortex. These originate mainly from pyramidal cells of layer V of the primary auditory cortex and terminate in regions of the IC which vary across species (for discussion about eutherians, see Aitkin 1990). Injections of retrograde tracer in the IC of opossums label neurons in layer V of the putatively auditory temporal cortex, with many more cells labeled on the ipsilateral than contralateral side (Willard and Martin 1983). Retrograde labeling of cells in the ipsilateral temporal neocortex has also been reported following IC tracer injection in quolls (Aitkin et al. 1986b). Within the IC of opossums, cortico-collicular fibers terminate most densely in DC, more lightly in EN, and not'at all in CNIC (Martin 1968; Willard and Martin 1983).

In addition to an auditory input provided from the CNIC (Kudo and Niimi 1980; Saldaña and Merchán 1992), the EN of eutherian species (Schroeder and Jane 1976; Aitkin et al. 1978b) receives somatosensory input from the dorsal column nuclei. More extensive studies in opossums have led RoBards to consider EN as the somesthetic tectum because of the widespread ascending and descending somatosensory input to this nucleus in this species (Martin et al. 1975; RoBards et al. 1976; RoBards 1979). However, efferents from the IC also terminate in EN of opossums (Jane et al. 1968). When the anatomical and physiological data are compared across species it is more likely that EN is a site of interaction between the auditory and somatosensory systems.

6.3.3 Efferents from the Inferior Colliculus

The question of efferent tectal pathways in a marsupial has been examined in studies by Martin (1969) and Jane et al. (1968). Both reported the degeneration patterns consequent to lesions made in various parts of the inferior or superior colliculi of opossums. Almost all of the efferent pathways described in eutherians have been confirmed in opossums, including those to the principal division of the medial geniculate body, to the superior colliculus and smaller midbrain structures, contributions to the tectopontine tract and the reticular formation, and descending fibers to the superior olivary nuclei. The efferent projections from the

IC in opossums to the different subdivisions of the MG will be considered further in Chapter 7.

6.4 Physiological Studies of the Auditory Midbrain

Although the physiological properties of single neurons in the auditory midbrain have been studied in considerable detail in eutherians (e.g., Aitkin 1986; Irvine 1992), there is a paucity of information about marsupials. However, data are available pertinent to frequency selectivity, binaural physiology, and the coding of space by units in the IC.

6.4.1 Tonotopic Organization of the Central Nucleus of the Inferior Colliculus and Frequency Selectivity of Single Units

In every mammal for which data are available, the CFs of units in the CNIC are arranged in an orderly fashion (Aitkin 1986), and marsupials are no exception to this general trend (Aitkin et al. 1978a, 1986b). Representative microelectrode penetrations through the IC of a brushtail possum (Fig. 6.4A) and Northern quoll (Fig. 6.4B) when reconstructed reveal an ascending sequence of CFs, indicating that the apex of the cochlea is represented in the dorsal part of the CNIC, whereas the base is represented in the ventral CNIC. In the quoll, units with CFs below 10kHz occupy a smaller volume of the CNIC than do those with higher CFs (Aitkin et al. 1986b).

It has been noted for a number of species that the isofrequency contours, imaginary lines constructed by joining together points having the same CF on a map of CNIC, have the same trajectory as do the dendritic laminae revealed by the Golgi technique (e.g., Aitkin 1986). The correspondence between anatomical and functional laminae in the CNIC is further strengthened in studies where functional labeling techniques are used. The substance 2-deoxyglucose (2DG) is taken up by respiring cells but does not complete a full metabolic cycle and so remains for a period of time in the tissue. If 2-DG is labeled with ^{14}C, the resultant molecule can be detected in tissue sections with autoradiography. When adult *Monodelphis* injected with 2-DG are exposed to pure tones, bands of autoradiographic labeling are observed in the inferior colliculus oriented from dorsomedial to ventrolateral (Reimer 1993). The orientation and trajectory of these bands are similar to the isofrequency contours observed in other species.

It has been suggested that the frequency selectivity of units in the IC of brushtail possums is generally less sharp than that of units in the IC of cats (Aitkin et al. 1978a). Tuning curves are very broad in the external nucleus – some span six octaves or more – and even in CNIC no extremely sharply tuned units are found. In association with broader tuning to tonal stimuli, units in the IC of possums are generally more strongly excited by broadband stimuli such as white noise than by spectrally simple stimuli such as tones (Aitkin et al. 1984). It is

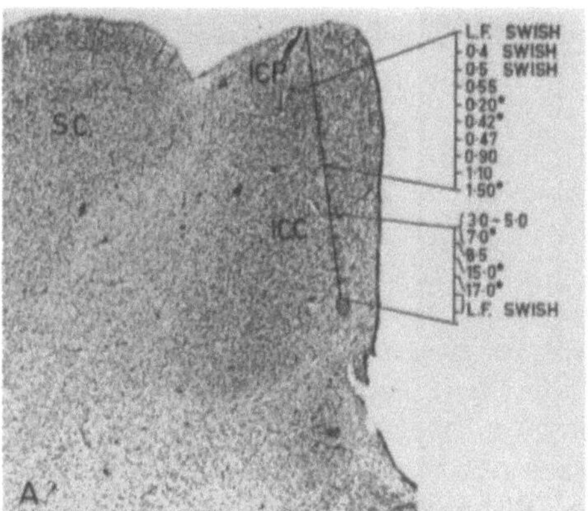

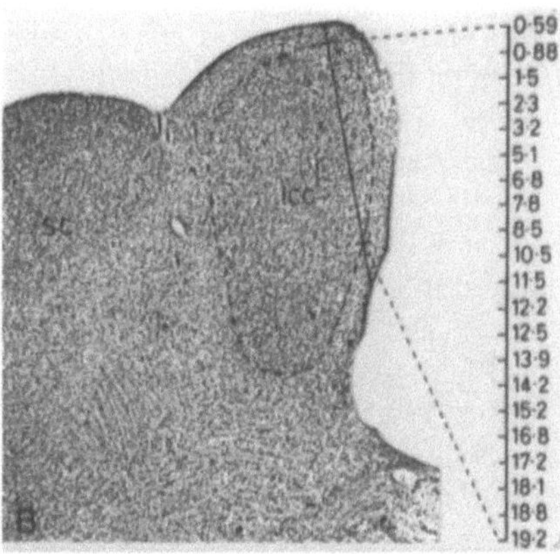

Fig. 6.4A,B. Sagittal sections (Nissl stain) through the midbrains of **A** brushtail possum and **B** Northern quoll. Microelectrode penetrations (*vertical lines through section*) recorded single and multiunit activity evoked by pure tone pips, mostly within the central nucleus of the inferior colliculus (*ICC*). The *dashed line* in **B** indicates the approximate border of *ICC*, separating it from the cortical regions (including the pericentral nucleus, *ICP*), and the superior colliculus (*SC*)

interesting that the vocalizations of brushtail possums are spectrally very broad (see Chap. 4) and would thus be potent activators of unit activity in the IC of this species.

6.4.2 Auditory Midbrain and Sound Localization

As indicated in Section 6.3.3, afferents reaching the auditory midbrain originate mainly from structures receiving binaural input. In the brushtail possum, 85% of units are binaurally influenced and three principal forms of binaural interaction have been recognized (Aitkin et al. 1978a). For one group, having low CFs, the discharge rate is influenced by the interaural time delay between binaural stimuli. For others, the response to binaural stimuli is an excitatory summation of the effects of stimulation of each ear alone. For a third group, stimuli applied to the contralateral ear are excitatory, whereas the same stimulus applied to the ipsilateral ear can suppress or inhibit the response to contralateral stimuli. Units in the third group usually have high CFs and are sensitive to the interaural intensity difference.

The localization of a sound source in space is made possible by the combination of neuronal sensitivity to interaural time and intensity differences (binaural cues) and the direction-dependent properties of the pinna (essentially a monaural cue) (for review, see e.g., Irvine 1986; Heffner and Masterton 1990). The operation of neural mechanisms in the auditory brainstem results in the emergence of neural sensitivity to the spatial location of a sound source. Neurons become spatially tuned and respond best to sounds located at a particular location in the horizontal (azimuthal) and vertical (elevational) planes. In the IC of cats and possums there are neurons classified as directionally sensitive, and for some of these units the same preferred azimuth is demonstrated over a range of sound-source intensities (Aitkin et al. 1984).

For example, plots of discharge strength (spike counts) against the azimuthal location of a sound source (where 0° is straight ahead) reveal for two units in the possum IC peaks in firing at particular azimuthal locations that vary little at four different sound levels for each unit (Fig. 6.5). For the unit in Fig. 6.5A the peak occurs at 40–50° contralateral to the midline, whereas for the lower unit it occurs at 30°. A population of similar units with differing azimuthal preferences could collectively encode the azimuthal location of any sound source.

For such a population coding strategy to operate, the angles made by the acoustical axes of the pinnae with the head and sound source need to be incorporated. For example, the response profile for another unit from the IC of a possum shows a minor peak in firing at about 60° contralateral to the midline and a steep decline in firing between 40° and 30° from the midline (Fig. 6.5C, filled circles). Such a unit could signal the location of a sound source by the change from zero

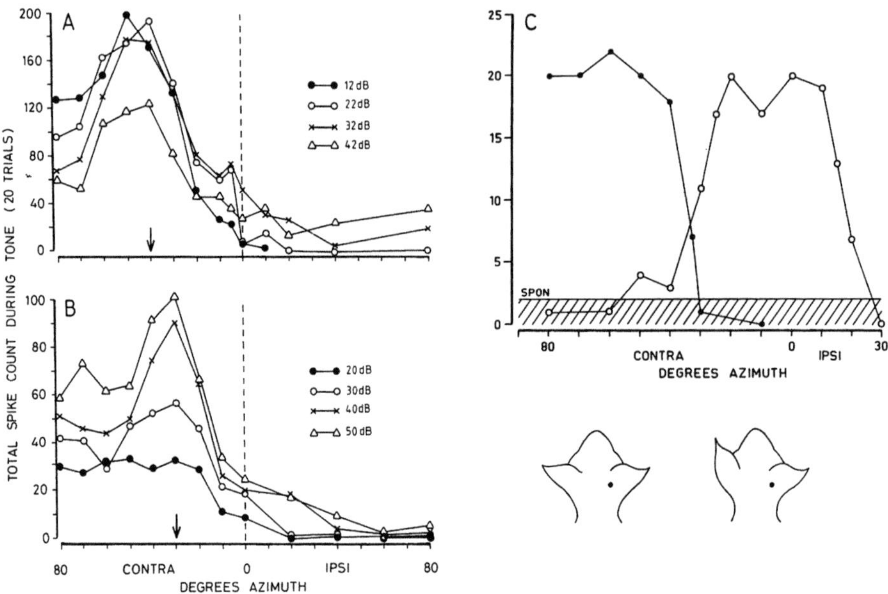

Fig. 6.5A–C. Coding of spatial location of a sound source by units in the inferior colliculus. Graphs relating the angular location of a sound source in the horizontal plane (*abscissa*: degrees azimuth) to discharge rate (*ordinate*: total spike count during 20 repetitions of a stimulus) for three units from the inferior colliculus of brushtail possums. **A,B** Two units stimulated by broadband noise show peak firing (*arrows*) at **A** 40° contralateral (*CONTRA*) or **B** 30°, irrespective of the sound level (dB), and very low firing rates ipsilaterally (*IPSI*). These units are thus selective for the spatial location of a noise stimulus. **C** Unit showing a steep decline in firing from an irregular plateau at 40° to 30° contra, at which azimuth firing is no greater than the level of spontaneous activity (*SPON*). When the contralateral pinna is bent through an angle of about 50° (*cartoon, below* **C**), the discharge border shifts correspondingly to ipsi azimuths, indicating the importance of the pinna in shaping the directionality of spatial coding. The *filled circles* in the cartoons indicate the side from which recordings were made

64

to near maximal firing, particularly if the source were moving. However, displacement of the left pinna (contralateral to the IC from which recordings were made) towards the right displaces the steep decline in firing by about the same amount. Thus the interpretation of the firing rate function is altered by alterations in pinna position.

The superior colliculus (SC) is as prominent as the IC in the midbrain of most marsupial species (Fig. 6.1). The deeper layers of the SC are important for the integration of auditory and somatosensory information with visual input and are concerned with the direction of gaze and the control of head orientation. Although primarily visual in function (see, e.g., Schiller 1984), the SC is involved with reflex head, pinna and eye orientation to sound (e.g., Stein and Clamann 1981). Neuronal activity in response to sounds located in different spatial regions has been investigated in the SC of the tammar wallaby (Withington et al. 1995). This study, utilizing noise-evoked multi-unit activity, characterized the directional properties of as many as seven loci within the SC in a given animal. Spatial tuning to sounds emanating from contralateral sound sources appeared sharper than that seen in the SC of eutherians. There was a "map" or gradient of sound-source location within the SC similar to that observed in eutherian species. Neurons in the rostral SC were tuned to sounds issuing from the anterior azimuthal field, whereas sounds originating posteriorly caused greatest activity in the caudal SC; intermediate sound-source locations correlated with greatest activity in the rostrocaudal center of the SC. This study also noted a "tight spatial alignment" between the best azimuthal position for visual responses in superficial layers of the SC and peak angle of auditory responses in deep layers.

6.4.3 Auditory Midbrain and Acoustic Reflexes

Pinna movements are one of a number of acousticomotor events important in hearing. Acoustic reflexes are clearly of great survival value since they can orient an animal towards the source of food or help avoid a potential predator. Some reflexes, including the middle ear muscle reflexes (Borg 1973) and startle responses (Davis et al. 1982), have such short latencies that they must involve very few relays between the ear and effector muscles and are unlikely to involve the auditory midbrain. However, it has long been known that interference with the inferior and superior colliculus alters pinna reflexes and the orientation movements of the head and eyes evoked by sound. Pinna movements can be evoked by electrical stimulation of the superior and inferior colliculus (Syka and Straschill 1970; Syka and Radil-Weiss 1971), but are not evoked by auditory stimuli after lesions of the superior colliculus (Sprague and Meikle 1965). Some of the pinna muscles of cats are innervated from a region in the motor nucleus of the facial nerve (Populin and Yin 1995) which receives input from a paralemniscal tegmental zone (Henkel and Edwards 1978). This region in turn receives descending projections from the superior colliculus (Henkel 1981). Part of the auditory input to the superior colliculus originates from the external nucleus of the inferior colliculus (see Aitkin 1986).

The role of the pinna in providing a direction-dependent amplification of sound has been discussed above and in Chapter 5. The number of different pinna movements and their importance vary across ethological and phylogenetic borders. Among marsupials, the pinnas of macropodids are remarkably mobile and can swing around to point rearwards, whereas those of the quoll have only a few configurations. The tips of the pinnas of striped possums and *Monodelphis* opossums exhibit fluttering, shivering movements (L. Aitkin, unpubl. observ.), rather reminiscent of the ear movements of certain echo-locating bats. It is not clear if these have an acoustic function or are stress responses. However, attempts have been made to record ultrasonic emissions from Monodelphis opossums, without success (B. Masterton, pers. comm.).

6.5 Future Research Directions

It is clear that there is as much diversity in the organization of the auditory pathway of marsupials as there is among eutherians. Substantive data are currently available only for three species, Virginia opossums, quolls, and brushtail possums. These are found on two continents and belong to three families. A proliferation of new knowledge is likely to arise from studies of the *Monodelphis* opossum, but there will still be large gaps in our understanding of the comparative anatomy of the auditory system in different marsupial families. Surveys need to be carried out comparable to those made in the visual system (e.g., Sanderson et al. 1979, 1984, 1987), in which anterograde tracer injections were made into the retinas of different marsupial species and the patterns of termination of retinogeniculate fibers examined. Thus, for example, we need to be able to examine the wiring of the lower auditory system in marsupials. Are there patterns common to all marsupials (equivalent to the marsupial cochlear nucleus cytoarchitectural patterns), to diprotodonts and to polyprotodonts, or to arboreal versus terrestrial carnivores, for example?

The handful of microelectrode studies in marsupials has produced some intriguing results that call for confirmation and extension, particularly in neuroethological directions, such as the relationships between vocal stimuli and neuronal responses and between sound localization and behavior. It would also be interesting to know if broadness of tuning relative to eutherians characterizes neurons in the lower auditory pathway (cochlear and superior olivary nuclei) as it appears to do in the midbrain? If so, could these features be explained at the level of the outer hair-cell tuning mechanism in the cochlea?

Chapter 7

Thalamocortical Auditory System

7.1 Medial Geniculate Body

7.1.1 Structure

Since the Golgi study of cats by Morest (1964), subdivision of the medial geniculate body (MG) into three major divisions is generally accepted for most species, although ideally this should be based on examination in Golgi material of dendritic arrangements. Morest and Winer (1986) stress the similarity between cats and opossums in the preservation of the major divisions of the MG when studied with Golgi material, although they indicate that boundaries between subdivisions in Nissl-stained material are less conspicuous in opossums than in cats. One difference they noted was the paucity of Golgi type II cells (short axon cells which act as local interneurons) in opossums. The MG of brushtail possums (Aitkin and Gates 1983) and Northern quolls (Kudo et al. 1989) has also been subdivided into ventral, medial, and dorsal divisions on the basis of packing density and dendritic labeling with horseradish peroxidase.

The ventral division of opossums occupies most of the ventral and lateral parts of the MG and is populated by densely packed, medium-sized cells. This division has two nuclei, the *pars ovoidea* (Fig. 7.1, OV) and *pars lateralis* (LV); cells in the latter in cats and opossums are arranged in rows with their dendrites forming laminae parallel to the lateral surface of the MG (Fig. 7.1), whereas cells in OV form curved or coiled arrangements (Morest 1965; Morest and Winer 1986). The ventrolateral nucleus (VL) is found immediately ventral to OV, and the medial division (Fig. 7.1, M) lies further medially. The latter is populated by large and small multipolar cells. The suprageniculate nucleus is positioned dorsal to M and consists of a dorsal (SGd) and ventral (SGv) subdivision. In Northern quolls a posterior nuclear group (Po) can be recognized rostral to SG (Kudo et al. 1989), part of which (lateral posterior nucleus, LPc) can be seen in Fig. 7.1, from an opossum. The analogous region in cats receives auditory input (Phillips and Irvine 1979). The dorsal division in opossums contains cells with radiating dendrites whose trajectories are flattened in the superficial dorsal nucleus and marginal zone (Fig. 7.1, DS, MZ) but are more open in the dorsal and deep dorsal nuclei (Fig. 7.1, D, DD). In cats the caudal tip of the MG is occupied entirely by the dorsal division, whereas in opossums the dorsal nucleus does not become distinct for several hundred micrometers and even then is comparatively reduced in size (Morest and Winer 1986). The suprageniculate nucleus receives auditory connections (Calford 1983), as does the nucleus of the brachium of the inferior colliculus (NBIC), further caudally (Aitkin and Jones 1992).

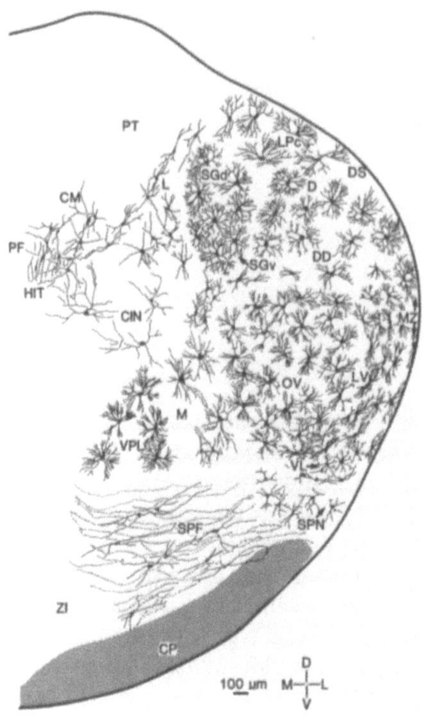

Fig. 7.1. Reconstruction in the transverse plane through posterior thalamus of a rapid Golgi impregnation from a 63-day opossum pouch young to show the dendritic arborizations of cells in the posterior thalamus. Auditory structures: *OV* pars ovoidea; *LV* pars lateralis; *VL* ventrolateral nucleus; *M* medial division; *SGd* suprageniculate nucleus, dorsal subdivision; *SGv* suprageniculate nucleus, ventral subdivision; *LPc* lateral posterior nucleus; *D* dorsal nuclei; *DD* deep dorsal nuclei; *DS* superficial dorsal nucleus; *MZ* marginal zone. Others: *CIN*, central intralaminar nucleus; *CM*, central medial nucleus; *CP*, cerebral peduncle; *HIT*, habenulo-interpeduncular tract; *PF*, parafascicular nucleus; *PT*, pretectum; *SPF*, subparafascicular nucleus; *SPN*, suprapeduncular nucleus; *VPL*, ventroposterolateral nucleus; *ZI*, zona incerta; *L*, posterior limitans nucleus (From Morest and Winer 1986, with permission of Springer-Verlag GmbH & Co. KG)

Tonotopic organization in the medial geniculate body of cats has been studied by a number of authors (Aitkin and Webster 1972; Calford and Webster 1981; Imig and Morel 1985). The most detailed of these, by Imig and Morel (1985), reveals the highly precise tonotopic organization of the ventral nucleus (V, the composite of OV and LV above), and proposes a model of the cochlear representation with a planar component (corresponding to LV) and a concentric component (corresponding to OV). In contrast, neurons in the dorsal and medial divisions are generally broadly tuned to tonal stimuli or respond only sluggishly to tones; tonotopic arrays in these regions, if they exist at all, are difficult to discern using microelectrode recording. No equivalent studies have been made of the organization of the MG in marsupials, although some idea of cochlear representation in the MG has been gained from studies of anatomical connections (see Sect. 7.4).

7.1.2 Afferent Connections

The axonal endings in the MG are also stained in rapid Golgi preparations (Morest and Winer 1986). A number of different axonal types can be recognized, some of which are extrinsic and some intrinsic to the MG. In LV many ascending axons (Fig. 7.2, 2) form a meshwork along the secondary dendrites (Fig. 7.2, stippling) of principal cells; some terminate as "nests" (Fig. 7.2, 5), whereas others are much finer (Fig. 7.2, 3) and run parallel to the dendrites.

The basic pattern of connections from midbrain structures in opossums is similar to that in cats, with afferent fibers of collicular origin forming very complex axonal nests with fibers of cortical and intrinsic origin in both species (Morest and Winer 1986). Lesions of the CNIC and DC lead to degeneration in the ventral and medial division, but not in the dorsal division, which occurs throughout the rostrocaudal extent of the MG. The dorsal division is supplied by fibers of the lateral tegmentum. Allowing for differences in nomenclature, a similar result was observed following lesions of the IC in the brushtail possum (marsupial phalanger; Rockel et al. 1972). In *Monodelphis* opossums, injections of anterograde tracers into CNIC provide label to the entire MG, with least density medially and rostrodorsally (Frost and Masterton 1993; Frost 1995). However, details about the specific interconnections of IC and MG subdivisions (e.g., Calford and Aitkin 1983) remain to be worked out for any marsupial. Telencephalic projections of the MG have been described in four marsupial species, the brushtail possum (Aitkin and Gates 1983), Virginia opossum (Kudo et al. 1986), Northern quoll (Kudo et al. 1989), and the South American gray short-tailed opossum (Frost and Masterton 1992, 1993). These will be discussed following an appraisal of auditory cortical fields, to which projections are directed.

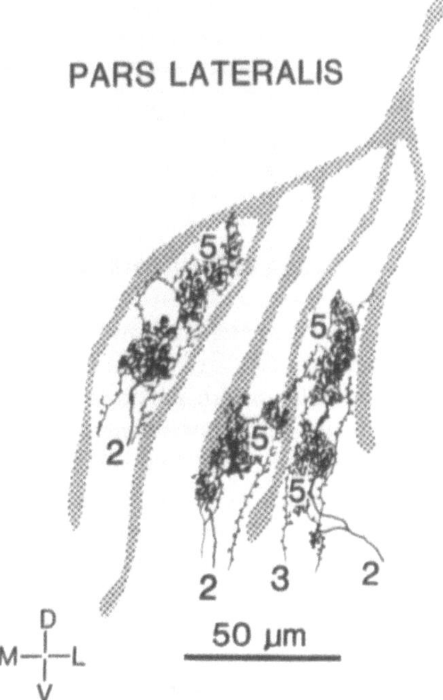

Fig. 7.2. Reconstruction of axonal plexes in a transverse section through *pars lateralis* of a 76-day-old opossum pouch young, stained with the rapid Golgi technique. *Numerals* identify axons having different morphologies, in relation to the secondary dendrites of a principal cell (*stippling*). (Morest and Winer 1986, with permission of Springer-Verlag)

7.1.3 Comparisons with Lateral Geniculate Nucleus (LG) of Marsupials

Sanderson and his colleagues (1978, 1984, 1987) have surveyed the organization of the lateral geniculate nucleus, the thalamic visual nucleus corresponding to the MG, in a variety of Australian marsupials by injecting anterograde tracer substances into the retina. They examined the pattern of retinogeniculate terminations in relation to the lamination of the LG. The LG of eutherians varies in complexity and number of laminae as a function of phylogenetic position (Sanderson 1986). For example, the LG of insectivores and most rodents is unlaminated and receives most input from the contralateral retina, whereas primates and carnivores have several layers, each receiving input from one eye or the other.

The LG of *D. marsupialis* shows no cytoarchitectonic lamination (Oswaldo-Cruz and Rocha-Miranda 1968), but injections of anterograde tracer into the retina provide a "quasi-laminar" arrangement of retinal endings (Lent et al. 1976). In contrast, no separation of inputs from each eye is demonstrable in *D. virginiana*, whereas there is considerable separation in the South American opossum *Marmosa mitis* (Royce et al. 1976). A similar variation is found in diprotodont marsupials, and the degree of lamination does not appear to correlate with habitat or nocturnal/diurnal activity patterns (Sanderson et al. 1987). Retinal afferents from one eye or the other are segregated in layers and the number of terminal bands vary from eight to eleven. The feathertail glider, noted in Chapter 6 for its unusual cochlear nucleus, is also unusual in having a poorly laminated LG and consequently considerable binocular overlap.

It would thus appear that the arrangements of laminae within the LG and the organization of retinal afferents to these laminae show more variation across marsupials that is not strictly phylogenetic than is the case with eutherians, but the laminated pattern would appear to have arisen within an ancestor common to both orders. More details of this kind are needed in the auditory system of marsupials.

7.2 Evolution of the Neocortex

The cerebral cortex of mammals has a highly uniform structure formed by layers of pyramidal cells (see Aitkin 1990, for review). The latter have roughly triangular cell bodies from which a long dendrite arises and extends, for most cells, towards the external surface of the neocortex. The different layers are the sites of termination of afferent fibers and sources of different efferent projections. Thus, for example, fibers from the medial geniculate body of cats terminate most densely in layer LV and to a small extent in layer I of the primary auditory cortex (Kelly and Wong 1981), whereas interhemispheric projections arise mostly from layer III (Imig and Brugge 1978).

The dorsal ventricular ridge (DVR) represents the highest synaptic level in reptiles for hearing and touch and shares with the dorsal pallium the representation of vision (Ulinski 1983; Frost and Masterton 1992). This tissue, however, lacks pyramidal cells and has its own distinctive architecture. The wulst of owls, the equivalent in certain avians to the visual cortex of eutherians (Pettigrew 1979), is also simpler in layering pattern than the mammalian neocortex, and lacks pyramidal cells. The DVR may have evolved from the dorsal telencephalon of Therapsid reptiles, and the differing arrangements of the cerebral cortex in modern reptiles and birds on the one hand and mammals on the other is an example of parallel evolution of functionally related structures.

The cerebral cortex varies in surface area partly as a function of body size but also in relation to phylogeny. Primates have a much greater amount of neocortex per unit of body weight, and humans have very much more than any other primate. The six-layered cerebral cortex is observed in all modern mammals, including monotremes. Abbie (1940) wrote: "The neocortex in the monotremes, while betraying some peculiarities, conforms to the general pattern common to all mammals" (p. 465). Descriptions of neocortical areas of marsupials (e.g., Rowe 1990) leave little doubt of gross similarities to eutherians. The auditory cortex of brushtail possums (Gates and Aitkin 1982) and Northern quolls (Aitkin et al. 1986b), defined by electrophysiological means (e.g., Fig. 7.3D,F), has six layers. Layer I is acellular, layer II (supragranular) is thin and composed of small darkly staining pyramidal cells with short apical dendrites. Layer III is much wider than II and contains a mixture of stellate and pyramidal cells, whereas layer IV (infragranular layer) is densely packed with small stellate cells. The borders of layer IV with layers III and V are not sharply defined. Layer V is as wide as layer III but has the largest cells (pyramidal) in the auditory cortex, and layer VI, also large, has mixed cell sizes and morphologies. The primary auditory cortex is clearly distinguishable from more dorsal cortex by virtue of the rather blurred lamination and wide infragranular layer of the auditory cortex. This pattern stops abruptly in the fundus of the rhinal sulcus to be replaced ventrally by the three-layered pattern of the entorhinal cortex, dominated by a thick, very dense layer II.

Northcutt and Kaas (1995) have recently reviewed the evolution of the mammalian neocortex by analyzing the way neocortical features compare between the closest relatives and, progressively, more distantly related taxa. In this way, features that change little over the course of evolution (primitive features) can be distinguished from those that become greatly modified (derived characters). On the basis of their findings, they reject the notion that neocortex evolves from the olfactory cortex (paleocortex) or hippocampus (archicortex), preferring the term "isocortex" to neocortex for this reason. They show that the isocortex has expanded a number of times; therefore many isocortical areas in living mammals are not homologous, in that their common ancestors are not likely to have possessed the multiplicity of, for example, visual areas that modern cats and monkeys possess. They conclude (p. 373) that "Although the exact origin of mammalian isocortex . . . is still disputed, it appears that the earliest mammals already had a six-layered isocortex with ten to 20 functional subdivisions."

Frost and Masterton (1992) discuss the evolution of the auditory cortex. They, too, compare the DVR of reptiles with the isocortex of mammals. The DVR receives fibers from the nuclei equivalent to the medial geniculate, ventrobasal complex and putamen (hearing, touch, and vision II) in a systematic fashion from medial to lateral. The visual geniculocortical pathway alone is represented in the

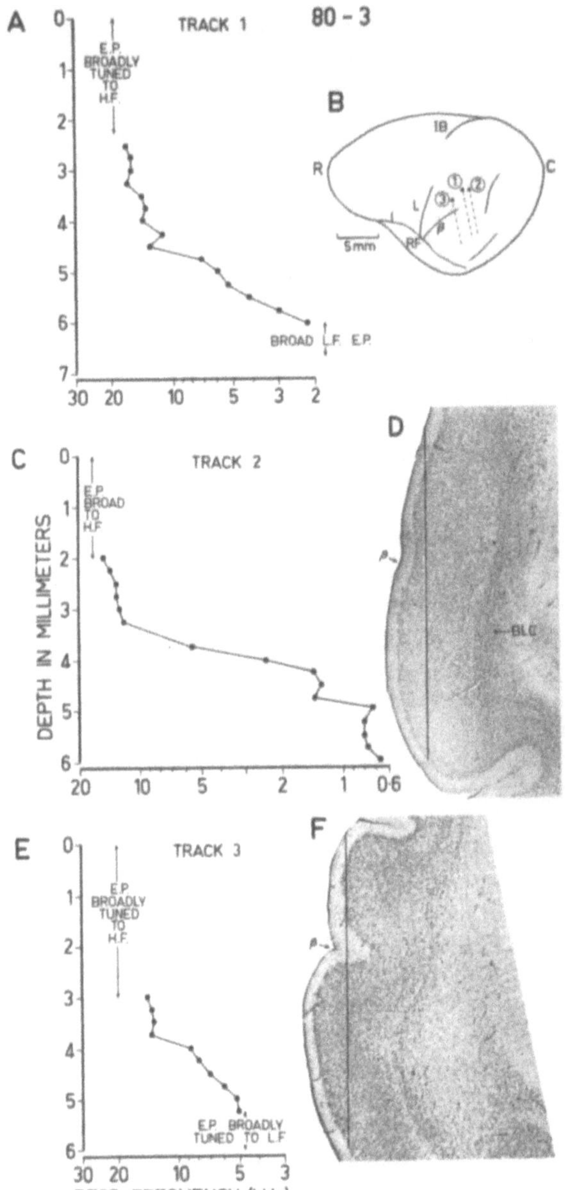

Fig. 7.3A–F. Reconstruction of three parallel dorsoventral microelectrode penetrations made along the rostrocaudal axis, the locations of which are shown by *number* on brushtail possum brain drawing in **B** (*C* caudal; *R* rostral; *L* labial sulcus; *RF* rhinal fissure; *IB* interbrachial sulcus). For all tracks (**A,C,E**) high frequencies (*H.F.*) were obtained dorsally and lower frequencies (*L.F.*) more ventrally. *E.P.* Evoked potential. Auditory (temporal) cortex lies ventral to a shallow sulcus labeled "β" in **B,D,F**. Frontal sections (**D,F**) show cortex from which recordings were made (*tracks 2 and 3*). *BLC* Basal lamina of the cerebral cortex. Calibration applies only to **B**; those for **D** and **F** are identical to the ordinates of **A,C,E**. (Gates and Aitkin 1982, with kind permission of Elsevier Science – NL, Sara Burgerhartstraat 25, 1055 KV Amsterdam, The Netherlands)

reptilian dorsal cortex. They suggest that during evolution the telencephalic sensory systems moved out onto the developing isocortex in an orderly way. They argue that the auditory system was the last relocated, to the temporal isocortex just dorsal to the olfactory cortex, above the rhinal fissure.

One way to test this hypothesis is to examine the thalamocortical auditory projections in mammals having the most remote common ancestry with placental mammals, such as monotremes and marsupials. No data are available for monotremes, but Kudo et al. (1986, 1989) and Frost and Masterton (1992, 1993) have used a variety of tracing techniques in three different marsupials to show that the caudal 30–50% of the MG does not project to the isocortex but instead to the lateral amygdala in the subcortical telencephalon. Mammals with a more recent divergence from the main line to primates show progressively less of the MG-to-lateral amygdala projection and more to the isocortex. Frost and Masterton (1992) consider that these results are "a series of snapshots of a transition or relocation of telencephalic auditory tissue" (p. 670). Further mention will be made of the significance of this major subcortical telencephalic projection zone when we consider the concept of auditory cortical fields and their number in marsupials.

7.3 Auditory Cortical Fields in Marsupials

Marsupials possess an auditory field analogous to the AI of eutherians in position, strength of response of constituent units, and completeness and orderliness of organization (see Chap. 2). Lende (1963) identified the auditory cortex in the opossum as a field lying above the posterior portion of the rhinal sulcus in which the largest evoked potentials to clicks clustered centrally. There was no indication of any other auditory field. Gates and Aitkin (1982) mapped the auditory cortex of the brushtail possum by making tangential microelectrode penetrations through the temporal cortex; their reconstructions suggest a single strong auditory representation (Fig. 7.3). Irrespective of whether penetrations were made rostrally (Fig. 7.3B,E) or caudally (Fig. 7.3B,C) through the auditory cortex, a descending sequence of CFs (best frequencies in Fig. 7.3) was encountered. This single auditory representation was flanked dorsally and ventrally by tissue from which only evoked potentials (E.P. in Fig. 7.3) in anesthetized animals were recorded, high dorsally and low ventrally. There is no suggestion of CF reversals or of other fields having sharply tuned neurons.

Very detailed maps have been published for the Northern quoll (Aitkin et al. 1986b); again, one main auditory field is described (Fig. 7.4). Microelectrode penetrations into the temporal cortex immediately dorsal to the rhinal fissure record sharply tuned responses to pure tone stimuli. The CFs of neurons encountered in the middle layers of the auditory cortex are high (>30 kHz) at the dorsorostral limit of the field and low (<2 kHz) ventrocaudally. Neuron CFs are arranged in a regular sequence such that roughly parallel isofrequency contours can be drawn across the cortex running in a dorsocaudal-to-ventrorostral direction (Fig. 7.4). Weak auditory drive characterizes the edges of this main field at most points, but a tendency for CFs to decrease beyond the rostral high CF border was noted in some experiments (Aitkin et al. 1986b), suggesting the existence of a second auditory field rostral to the first. The latter, because of its size, strength of responses, and sharpness of tuning of its component elements, was considered to be homologous with the AI of eutherians.

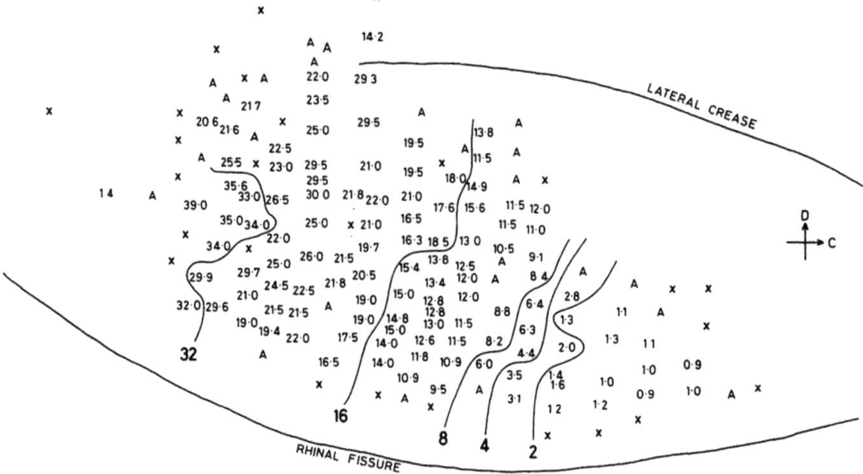

Fig. 7.4. The distribution of CFs across the acoustically responsive cortical surface in a Northern quoll. Sites marked with an *A* responded to auditory stimuli but no CF could be assigned; those indicated by *X* were unresponsive to tones. *Lines* identify estimated isofrequency contours in one-octave intervals from 2 to 32 kHz. *D* Dorsal; *C* caudal. (Modified from Aitkin et al. 1986b, with permission)

Krubitzer (1995) has examined the organization of sensory cortex across species. She finds differences in the amount of cortex devoted to each sensory system, the number of fields within that sensory cortex, the size and configuration of individual fields, and the patterns of connections of homologous fields. The apparent restriction in the amount of auditory cortex in marsupials could relate to a relative unimportance of hearing to the few marsupial species so studied, but this seems unlikely given the very vocal behavior of the brushtail possum and the acoustic needs imposed upon a nocturnal hunting lifestyle of an animal like the quoll. Another possible reason for the different organization of the auditory cortex may be that certain features of brain organization have evolved differently in marsupials compared with eutherians.

Merzenich and Schreiner (1992) imply the latter by referring to the major projection from the caudal part of the medial geniculate to the subcortical lateral amygdala, discussed above, that is present in at least three marsupials but as yet undescribed in eutherians. Perhaps the auditory representation in the lateral amygdala of marsupials may be the analog of a secondary auditory field in eutherians. In cats field AII lies ventral to AI, whereas the rhinal fissure and the olfactory cortex lie ventral to the AI of marsupials. The lateral amygdala, however, is adjacent medially to the auditory cortex, and it is reasonable to imagine that the equivalent area, in a common ancestor to eutherians and marsupials, may have migrated on to the cortical surface to form a new auditory cortical field. It would be very interesting to examine the auditory properties of neurons in the lateral amygdala of marsupials.

7.4 Auditory Cortical Connections of Marsupials

Aitkin and Gates (1983) recorded auditory-evoked potentials from the auditory cortex of brushtail possums prior to making small electrophoretic injections of retrograde tracers. As mentioned above, only one fully organized auditory cortical field was identified in this species, and afferent projections arise from all parts of the medial geniculate body. Those originating from the lateral sector are topographically organized, such that injections of tracer at low CF regions of the auditory cortex label neurons in the lateral part of the MG, whereas injections at higher CF loci label cells in more medial parts of the MG. Such projection patterns suggest that the auditory cortex of brushtail possums is analogous to the AI of eutherians.

Interhemispheric connections were also demonstrated in this study. It has long been known that marsupials lack the *corpus callosum*, the distinctive band of fibers which connects the two cerebral hemispheres in eutherians; instead the hemispheres communicate through the anterior commissure and the *fasciculus aberrans* (Abbie 1939; Heath and Jones 1971). In the brushtail possum neurons were labeled in the opposite hemisphere at positions corresponding to sites where large injections were made in physiologically characterized auditory cortex. Small electrophoretic injections did not provide label in the opposite hemisphere (Aitkin and Gates 1983). In the visual systems of brushtail possums and wallabies, interhemispheric projections are also essentially equivalent to those in the corpus callosum-possessing eutherians (Crewther et al. 1984; Sheng et al. 1990).

Kudo and his colleagues (1989), working with Northern quolls, made injections of wheat-germ agglutinan conjugated with horseradish peroxidase (WGA-HRP), a tracer which is transported in both anterograde and retrograde directions along axons, into the auditory cortex, defined after physiological mapping. Labeled cells and terminals were identified throughout the MG ipsilaterally and in three telencephalic areas, the contralateral and ipsilateral lateral amygdala, and the contralateral temporal cortex. They showed that the dorsal part of the lateral amygdala, which receives ascending input from the caudal part of the MG, also has reciprocal connections with the auditory cortex. This study is important because it shows clearly that the auditory cortex is *reciprocally* connected with all regions to which it is related – thalamus, subcortical telencephalon, and neocortex – in a manner identical to eutherians so studied, yet the *specific regions* receiving and giving connections are different from eutherians.

A substantial amount of information on the projections of posterior thalamic nuclei to the neocortex in the brushtail possum has been obtained by making injections of retrograde tracer in regions of the neocortex defined by positional rather than physiological criteria (e.g., Haight and Neylon 1978, 1979; Neylon and Haight 1983). In relation to the auditory system, Neylon and Haight (1983) used cytoarchitectural criteria for the auditory cortex of possums, derived from physiological experiments of Gates and Aitkin (1982), to locate their injections of tracer into the auditory cortex. They found that, in addition to the projection from MG, inputs arose from adjacent thalamic nuclei, the posterior group, PO, and the suprageniculate nucleus (SG).

According to Neylon and Haight (1983), projections from the SG to the auditory cortex have not been reported in eutherians, and they may only be minor in the Northern quoll (Kudo et al. 1989). However, responses to sound have been reported from neurons in the SG of cats (Calford 1983), and it may be difficult to compare regions so defined in different species. Projections from the SG to the temporal (presumed auditory) neocortex were also noted for the Virginia opossum by Kudo et al. (1986). This study additionally confirmed findings for the opossum, made in a retrograde degeneration study by Ravizza and Masterton (1972) and anterograde degeneration study of Ebner (1969), that the MG projects to a region of temporal neocortex where Lende (1963) mapped responses evoked by sound.

7.5 Effects of Lesions of Cortical Areas on Behavior

The functional organization of the auditory pathway of animals other than humans has been studied by making lesions in selected parts of the auditory system and observing the alterations in behavior that follow. Much of our knowledge has been obtained from work carried out in cats by Neff (e.g., Neff 1961) and, more recently, Masterton and his colleagues (e.g., Jenkins and Masterton 1982; Masterton et al. 1992, 1994). Much attention has been given to the auditory cortex, partly because surgical intervention in this region of the auditory pathway is simpler than in any other (Heffner and Masterton 1975; Jenkins and Merzenich 1984). One shortcoming that is consistently associated with a lesion of the auditory cortex is a deficit in localizing sounds at azimuthal locations contralateral to the lesion (see Aitkin 1990, for review).

Ravizza and Masterton (1971, 1972) examined the deficits in reflex and conditioned behavior after ablation of most or all of the neocortex of Virginia opossums. Complete decortication ensured that all of the auditory cortex had been removed; this was confirmed postmortem by the demonstration of complete retrograde degeneration of neurons within the MG bilaterally (Ravizza and Masterton 1972). These animals could identify a change in the direction from which a sound is coming but they showed deficits in their abilities to distinguish between azimuthal locations, with a threshold rise from about 5° to 24° after decortication. Other hearing abilities – vertical localization, intensity, and frequency sensitivity – were not obviously altered by neocortical ablation in opossums. In this regard, opossums behave in the same way as do all studied eutherians following a lesion of the auditory cortex.

7.6 Future Research Directions

The functional organization of the neocortex has been used by many authors as an index of the "primitiveness" of a species. There are apparent differences between the few marsupials studied and those eutherians studied more completely, in the number of auditory fields, their arrangements, and their connections. The extent to which these differences are phylogenetic rather than ecological can only be decided by surveying a larger number of marsupials occupying different ecological niches. The answers could cast light on the evolution of the brain.

More specifically, physiological studies of auditory cortical areas are singularly lacking in marsupials. Is there a strong binaural representation in the primary auditory cortex as there is in well-studied eutherians, such as cats? How do neurons in the auditory cortex respond to species-specific vocalizations? Is there physiological evidence for auditory fields outside the primary area, and does the lateral amygdala act as a neocortical field? Finally, what kind of auditory information is transferred between the two hemispheres, and does the lack of a corpus callosum deny marsupials perceptual abilities available to eutherians?

Chapter 8

Development of the Auditory System

8.1 Why Marsupials Are Important to the Developmental Biologist

Much of the "fetal" development of a marsupial occurs outside the uterus, often in a pouch (Latin *marsupium*). When it emerges from the birth canal the marsupial embryo has little nervous or muscular tissue to guide it to the mother's nipple; how it does so is still one of the major mysteries in biology, although it would seem that different strategies are used by different marsupial species (e.g., Tyndale-Biscoe 1973; Lee and Cockburn 1985). The newborn marsupial is deaf and blind, apparently insensitive to nociceptive stimuli, and only the chemical senses (and perhaps parts of the trigeminal system) may provide some sensory input. Thus, the newborn marsupial is in effect an externalized mammalian embryo whose subsequent development can be studied without significantly affecting the mother. Furthermore, structures in the newborn marsupial can be manipulated prior to the growth of nervous connections to many parts of the central nervous system, so that it is possible to examine how neural input influences both organogenesis and the normal pattern of connectivity in the brain. This chapter examines how the auditory system, particularly the neural components of this system, develops in marsupials. Although this subject caught the attention of auditory physiologists more than 50 years ago (McCrady et al. 1937, 1940), little research has been done on marsupial auditory development. This is surprising, given the accessibility of neonatal marsupials at such an early age, although this deficiency is now being rectified with the establishment of many colonies of the South American opossum *Monodelphis domestica*, which can be readily reared in captivity.

8.2 Embryogenesis of the Auditory System and Growth of Young Marsupials

McCrady (1938) has described in detail the embryology of the Virginia opossum prior to birth. The young develop in the uterus for 13 days after fertilization, and lactation lasts a further 70–80 days. The acoustic placode, from which the inner ear will develop, is recognizable on day 9; at this time the 3 primary divisions of the brain are recognizable. The forebrain consists only of its diencephalic portion, and the midbrain is demarcated laterally by two very shallow grooves. The acousticofacial ganglion is recognizable at day 9, with many cells being contributed by neural crest proliferation. At this time also those structures important in the sucking and breathing mechanisms show accelerated development relative to eutherians.

By day 11 all cranial nerves are present. The dendrites of the eighth nerve from the otocyst are separable into two groups, and the spiral ganglion is present. On day 12 the posterior semicircular canal has been formed from the otocyst, and all cranial nerves have reached their end organs. The lateral canal appears on day 13, the day on which the embryo is born. The vestibular system is still so undeveloped that the progress of the young from birth canal to nipple must occur without any assistance from that system.

McCrady (1938, p. 193) describes the state of the ear at birth:

"The cochlea duct . . . has grown out to about one-half turn. It contains no organ of Corti, tectorial membrane, scala tympani or scala vestibuli. The external auditory meatus is completely plugged with epitrichial cells, and the latter also overlie the developing pinna. . . . No nerve endings are present in either the acoustic or vestibular part of the inner ear."

J. Nelson (unpubl. observ.) has reviewed the growth of young *Dasyurus hallucatus* after birth (pouch young). Births occur around the shortest day of the year after a gestation period of about 25 days. There is rapid growth for the first 3 days after the young appear in the pouch, and the volume of milk in the gut is large at this time. There is a slower growth rate over the next 10 days (e.g., 8 d, Fig. 8.1), but in the subsequent 2 weeks the hind limbs begin to develop claws and the tail elongates; this period is associated with neurogenesis, the rapid growth of the brain and the first connections of the telencephalon.

By 45–50 days hair appears on the body (Fig. 8.1, 45 days), and the young have been able to move from one nipple to another; they sometimes are found outside the pouch. They are very mobile by 64 days (Fig. 8.1), and at around 72–76 days the young open their eyes (Fig. 8.1). Teeth have begun to erupt by day 85 (at which time the brain has an adult appearance), and the young make brief excursions out of the pouch and feed on soft foods. They are mostly weaned by 100 days and can kill prey at 120 days. Their time span as pouch young would thus be approximately 100 days.

In contrast, *M. domestica* breeds throughout the year but is a pouchless genus; neonates often cling to the mother's back (Fadem et al. 1982). Gestation is 15 days and the young remain attached to the nipples for 14 days after which they continue to feed but move around on and off nipples. They begin to eat solid food at about 30–35 days and are weaned by 60 days.

Developmental stages relevant to hearing in the tammar wallaby (*Macropus eugenii*) include 26–28 days of intrauterine development and a subsequent 250 days before the young leaves the pouch permanently, the opening of the external auditory meatus at 125–130 days postnatal, and eye opening at 140 days postnatal (Gummer and Mark 1994).

8.3 Physiological Measures of the Onset of Hearing in Marsupials

The onset of hearing has been studied in quolls by measuring startle responses to sound. Startle responses to noise bursts of 105 dB sound-pressure level first ap-

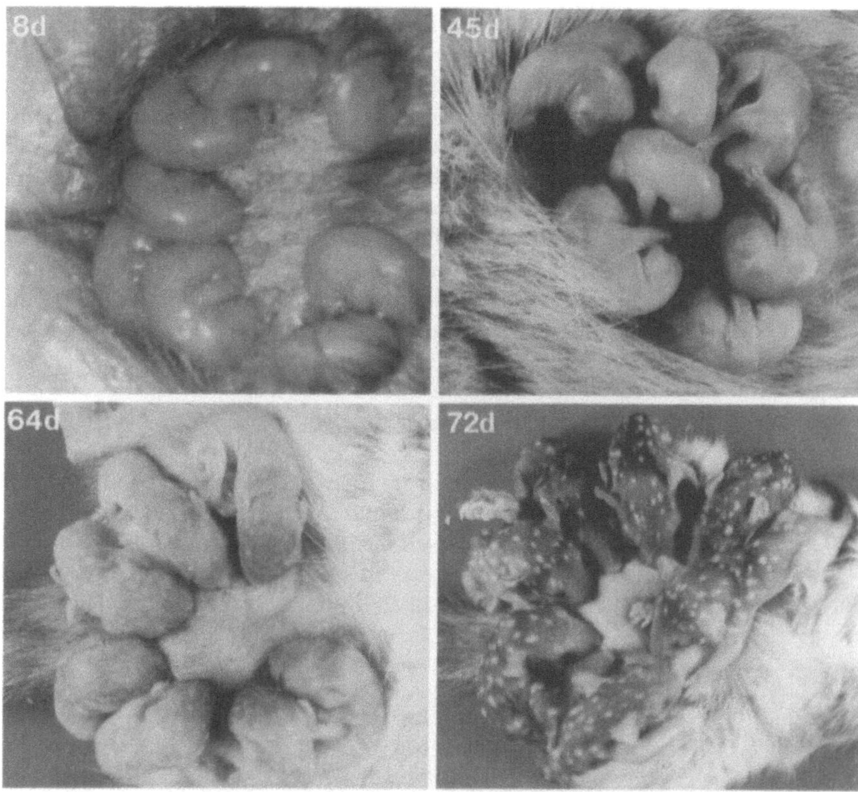

Fig. 8.1. Northern quoll pouch young at different ages, photographed at different magnifications. At *8 days* only the mouth and grasping forepaws are in any way mature. By *45 days* (*d*) hair has begun to erupt and hind limbs and tail are fully mobile. At *64 days* the skin pigmentation assumes the mottled appearance of the adult, and the pinna is mobile and adultlike. At *72 days* the dark fur of the adult is achieved, the eyes are beginning to open, and the animal frequently leaves the pouch between feeds. (Photographs kindly supplied by Dr. John Nelson)

pear at 60 days after arrival in the pouch, but only to occasional stimuli; forelimb rather than whole-body twitches were evoked (Aitkin et al. 1996a). The latter are elicited regularly at 67 days onward. For quolls, therefore, hearing begins about 98 days after conception and about 30 days before weaning. Unpublished observations in *Monodelphis* by F. Willard suggest that startle responses are evoked in this marsupial at 30 days, approximately 45 days after conception.

McCrady and his colleagues (1937, 1940) recorded the cochlear microphonic potential (CM) in the developing Virginia opossum. Although the presence of a CM does.not of itself indicate that the central auditory system is mature, the timing of maturation of hair-cell function and central transmission are closely related (Pujol and Hilding 1973). Cochlear potentials were first demonstrable in pouch-young aged 48 days (McCrady et al. 1940), when they were elicited by tones of mid-frequency range (2–6 kHz) at about 100 dB. A whole-body reflex response to sound was observed in pouch young of about this age (McCrady et al. 1937).

More accurate measures of the onset of auditory function are obtained by using auditory brain-stem responses recorded from the scalp and averaged for many repetitions of an auditory stimulus. In 68-day quolls, responses were elicited between 1–16 kHz with thresholds in excess of 55 dB. At 81–88 days responses occurred over the adult range at lower thresholds than observed in the adult (Fig. 8.2). Potentials evoked by clicks (Fig. 8.3, insets) recorded in *Monodelphis* from the skull immediately dorsal to the inferior colliculus show a similar complex waveform to adult ABR waves and exhibit a prominent peak at a latency of 2.5 ms that may correspond to the IC (see Chap. 3). Thresholds for visual detection of this average wave, summed for 500 stimuli, drop steadily as a function of age, from 83 dB at 24 days (the earliest age at which potential were evoked by clicks in this study) to 58 dB at 40 days (Fig. 8.3). In another litter, evoked potentials were not recorded at 26 days but were elicited at 28 days. It is likely that factors other than age (e.g., maternal nutrition) may influence the onset of hearing in this species.

Reimer (1996) recorded ABR waves from the scalp of developing *Monodelphis* opossums in response to pure tones. She found that consistent responses to pure tones were first obtained at 29 days, when thresholds to 8 and 12 kHz were about 60 dB. Given the above click data, it is easy to see that the first evoked potentials to transient stimuli such as clicks are likely to be recordable some days earlier

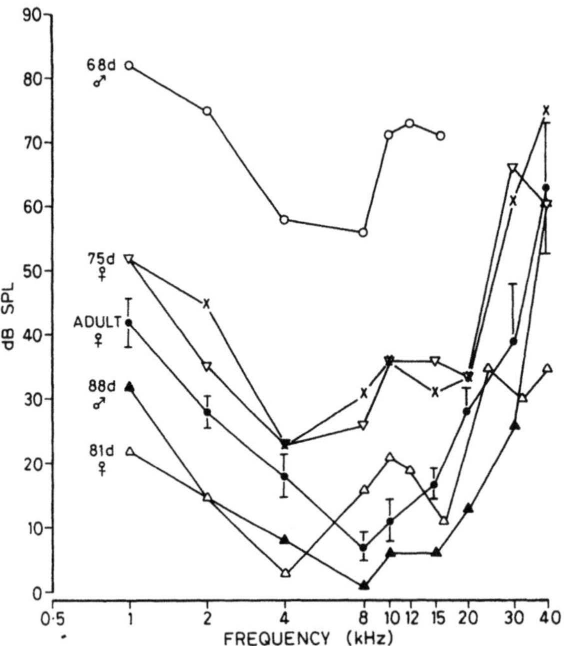

Fig. 8.2. ABR audiograms recorded from quolls of different ages. The auditory brain-stem response is measured from the vertex of the skull in reference to an electrode placed near the ear on the side to which sounds are presented. The threshold in dB at which the waveform is just visible above the noise is plotted against frequency in kHz. (Aitkin et al. 1996a, with permission of Wiley-Liss, Inc.)

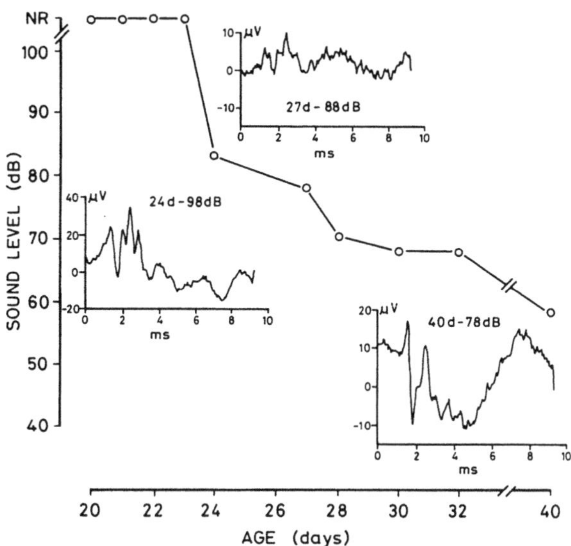

Fig. 8.3. *Graph* Threshold sound level (dB) for ABR waveform to click stimuli recorded from the skull immediately above the inferior colliculus of neonatal *Monodelphis domestica*, plotted against age in days. Prior to 24 days of age, no response can be recorded. *Insets* Samples of ABRs to clicks at the stated intensities and days (*d*). (L. Aitkin, S. Cochran, and S. Frost, unpubl. observ.)

than 29 days, several days before whole-body startle responses occur. As the animal ages, the range of effective frequencies broadens and the thresholds at these frequencies drop steadily, and an adult ABR audiogram is obtained at day 40 (Reimer 1996).

The relationship between the ABR and evoked potentials from auditory nuclei has been studied in pouch young of the tammar wallaby (*Macropus eugenii*) aged 90–250 days (Liu et al. 1996). There is a sequence in development from responses in the auditory nerve at day 101 to the IC at 116 days. In the latter structure, responses first appeared in its rostral pole, then caudally, and by day 125 the entire IC was responsive. These results make the interpretation of ABRs as indices of the onset of hearing problematic. Which component needs to be evoked in order for one to say that the animal *hears*? Perhaps the entire ABR sequence. In tammar wallabies that sequence seems to be established between 150–182 days (Liu et al. 1996).

In eutherians, brain-stem auditory evoked potentials have been recorded in fetal sheep after 117 d gestation (normal term about 147 days; Cook et al. 1987), in guinea pigs 12–15 days before term (63 d), but in most laboratory animals at, or some days after, birth (Pujol and Hilding 1973), and in ferrets as late as 32 days postnatal (Moore 1982). Eye blinks can be recorded as startle responses to vibroacoustic stimuli in human fetuses about two-thirds of the way through fetal life (Birnholz and Benacerraf 1983), although such a stimulus may activate cutaneous, as well as auditory, receptors.

Given the greater exposure of the marsupial fetus to environmental stimuli compared with eutherians, it is interesting to speculate about the precocity of marsupial hearing. Assuming, for *Didelphis* opossums, 13 days of embryogenesis in utero, a further 50 days in the pouch to the onset of startle responses to sound, and final departure from the pouch (weaning) at 85 days (McCrady 1938), hearing begins in this species about 64% of the way through "fetal" life. Hearing in *Monodelphis* opossums occurs at 24–30 days (depending on criteria), gestation at 15 days and weaning at 60 days; the onset of hearing in this species could be said to occur 56% of the way through fetal life. Hearing in quolls occurs rather later during development – startle responses at day 67, after a gestation period of 25 days and weaning at 100 days (74% through fetal life).

How does this compare with eutherians? Startle responses have been used to assess hearing in many species. Ferrets respond to loud claps and finger snaps at 32 days postnatal (Moore 1982) and mink at 29 days (Foss and Flottorp 1974); these mustelids appear to be one extreme, in the published literature, in terms of postconceptional age for the onset of hearing. Mice (9–14 days depending on species; Alford and Ruben 1963; Hack 1968; Ehret 1976) and rats (10 days; Wada 1923) are also altricial. At the other extreme, guinea pigs, with a gestation period of about 67 days, probably respond to sound in utero (Pujol and Hilding 1973). Brain-stem auditory evoked potentials have been recorded from sheep fetuses at 117 d gestation (normal term 147 days; Cook et al. 1987), indicating that sheep, like guinea pigs, are particularly precocial in respect to the development of hearing. Using birth rather than weaning as the staging index for eutherians, hearing could be said to occur 79% through fetal life in sheep and 76% in guinea pigs. On this basis, hearing is precociously developed in opossums but not in quolls. There may be a spectrum among marsupials as there is among eutherians in terms of the time of hearing onset.

8.4 Development of the Auditory Periphery

Changes in the outer ear during pouch life are shown for quolls (Fig. 8.4). From about 20 days a series of folds appears bilaterally below the occiput. At 33 days (Fig. 8.4, 33 days, left) the pinna is still attached to the scalp on all sides with its tip (arrow) pointing caudally. When the flap is removed and the developing eardrum exposed (arrow, Fig. 8.4, 33 days, right), the latter is thick and contiguous with connective tissue beneath. No middle ear cavity is visible at this stage.

At 45 days the folds that become the helix and tragus are present, and a tiny blind opening (arrow) can be seen ventral to the developing tragus (Fig. 8.4, 45 days). This opening remains blind at 51 days (Fig. 8.4, 51 days), although the middle ear cavity is largely free of fluid, and the ear drum has thinned and become more transparent. The pinna has developed the more pointed appearance of the adult pinna, but remains up to and including this age largely flattened to the scalp.

By 63 days the ear canal has opened (Fig. 8.4, 63 days, left), leading to the tympanic bone (arrowheads), which supports a transparent, shiny eardrum, through which the handle of the malleus is visible (white arrow, Fig. 8.4, 63 days, right). The middle ear cavity behind the eardrum is air-filled, and the outer

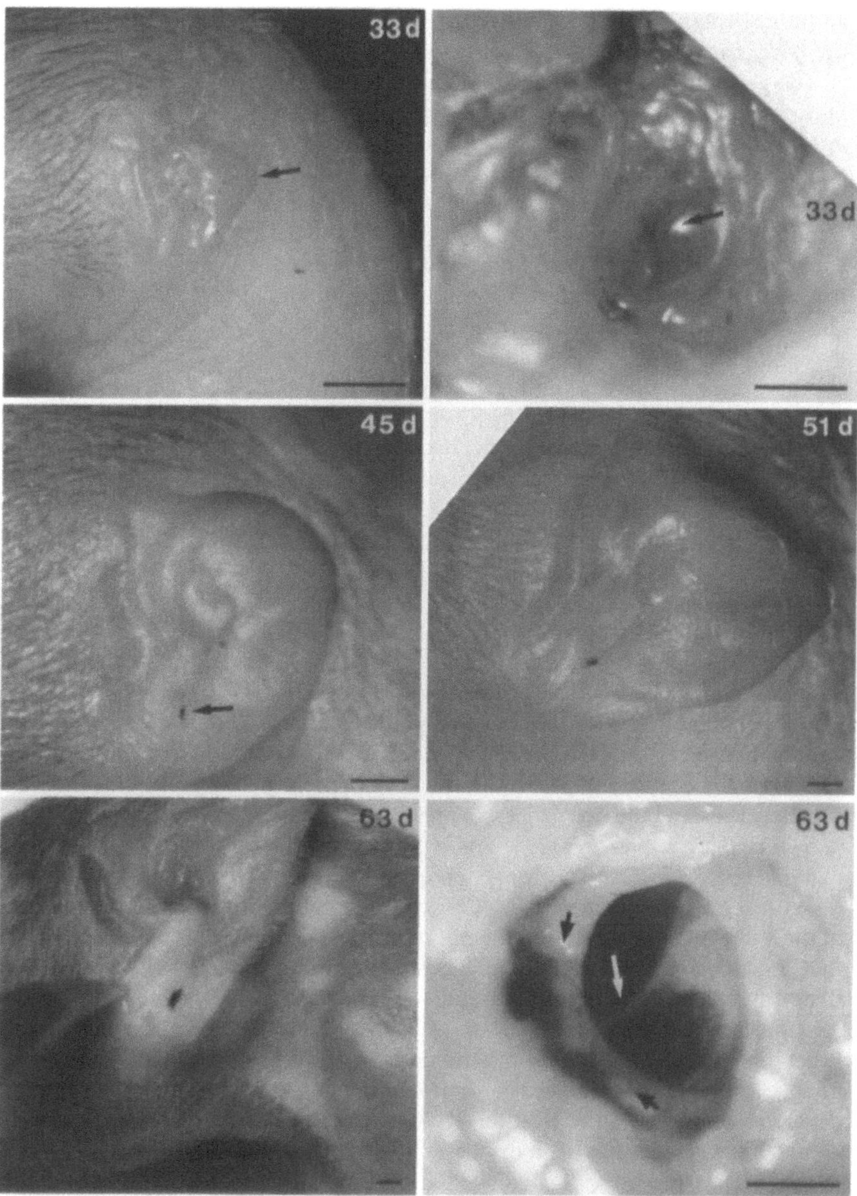

Fig. 8.4. Changes in the pinna and eardrum of Northern quolls as a function of age. In all frames, rostral to *left*, dorsal *up*. Calibration 1 mm. The *curved object* in *lower left frame* is a hemostat used to pull away the ventral tissue of the pinna to expose the open ear canal at *63 d* (days) *Arrows*, *33 d left*, the tip of the developing pinna; *33 d right*, the eardrum visible after dissection; *45 d left*, a blind opening that will become the external ear canal; *63 d right*, the manubrium of the malleus (*white*) and the tympanic bone (*black*). (Aitkin et al. 1996b, with permission of Wiley-Liss, Inc.)

auditory system would be receptive to sound albeit at a reduced pressure due to the fairly narrow outer ear canal.

The general trend of the morphological development of the cochlea has long been recognized as progressing from the base (the end nearest the stapes) to the apex. Interestingly it has been shown for a number of species that physiological responses to sound first occur to frequencies in the middle range of the species' audiogram and not in regions corresponding to the base (Rubel 1978). This is not a contradiction if the mechanical properties of the base as it first develops correspond to frequencies that are, in the adult, of the middle range. Such a hypothesis has been suggested in a different form by Rubel and Ryals (1983) for the maturing chicken basilar papilla.

Tuning curves are shallow and insensitive at the onset of hearing and become sharper, more adultlike after a few weeks of life (Aitkin and Moore 1975; Carlier et al. 1979; Romand 1983). During this period the output of the hair cells is related directly to the incoming vibrations, because outer hair cell and efferent connectivity have yet to mature. In other words the cochlea is still a passive device, active properties not having yet developed (see Chap. 5).

8.4.1 Mechanical Structures

Saunders and his colleagues (1993) have reviewed the contribution of middle-ear sound conduction to auditory development. For some species there is a striking parallel between the development of middle ear transmission and threshold sensitivity of the cochlea, with middle ear development preceding cochlea maturation in gerbils and the two processes maturing at about the same time in mice and hamsters. Saunders et al. (1993) argue that maturation of the conductive system in some species is a rate-limiting step in auditory development.

With regard to marsupials, McCrady (1938) commented on the development of the auditory ossicles and their embryological origin in *D. virginiana*. He noted that the anterior process of the malleus extends in a semicircle halfway around the tympanic annulus (tympanic ring, see Chap. 5) and discussed the possible relationships between the quadrate of reptiles and incus of mammals and between the articular of reptiles and the malleus of mammals. McCrady (1938) was unable to find a clear connection between the stapes and the hyoid arch in his developmental series; the relationship between the stapes of mammals and the columella of reptiles has been discussed in Chapter 5. Middle-ear ontogeny in *M. domestica* has been investigated by Filan (1991), but this study was directed more towards a better understanding of the immediate consequences of suckling in a neonate marsupial and the epigenetic factors that constrain morphogenesis rather than the development of hearing. Of relevance to middle ear processes is the observation that developing *Monodelphis* have jaw articulations which are neither mammalian (dentary-squamosal) nor reptilian (quadrate-articular). The tensor tympani muscle is well developed by day 3 and the stapedius muscle by day 5. The malleus and incus also develop more quickly than the stapes.

The basic differentiation of acousticomechanical structures in the inner ear of developing marsupials has been studied in the Northern quoll by Gemmel and Nelson (1992) and in *Monodelphis* by Willard (1993). At birth, when the quoll first appears in the pouch, the cochlear enlargement is distinguishable from the vestibular sac. The cochlea is initially a straight tube (day 3), the tip turns in a caudolateral direction and forms a coil (day 6), there is increasing coiling of the cochlea (day 9), and by day 11 there are two complete coils. At day 25, when 2.5 coils are present, the organ of Corti begins to differentiate, and the stapes becomes attached to the oval window. At day 33 there is still much fibrous tissue within the cochlear duct, and the tectorial membrane appears flattened due to the presence of cells within the internal sulcus and tunnel of Corti. By day 37 the tectorial membrane is free, due to resorption of these cells, and by day 50 the organ of Corti is fully mature. At day 55 the external ear duct opens. By day 80 the quoll spends much of its time out of the pouch and has been weaned.

Events are appropriately earlier in *Monodelphis*, which is weaned at about 60 days. At postnatal day 0 (PND 0) the otocyst has three-quarters of a turn, and there is no penetration of auditory dendrites (Willard 1993). By PND 6 there are 1.5 turns, and the appearance is equivalent to ED 16 in the mouse; by PND 8, 1.75 turns are present, and by day 14 the complete spiral is formed but hair cells remain immature. The cochlea is equivalent to that of a mouse at PND 3–4.

These observations confirm those made by others who have studied the general embryology of the developing marsupial (McCrady 1938; Hill and Hill 1955). The relationship between the developing sensory epithelia and the onset of hearing in eutherian species has been given considerable attention in recent years.

8.4.2 Ultrastructural Development of the Cochlea in Eutherians

Pujol and his colleagues (Pujol 1985; Pujol et al. 1997) have reviewed the development of hair cells and synapses at the bases of the hair cells in eutherian cochleae. It is not clear when inner (IHC) and outer (OHC) hair cells become distinguishable. There are already microvilli and kinocilia present before differentiation, and some nerve fibers are invading the epithelium. This could indicate that some of these histologically indistinguishable cells have already begun to differentiate as sensory cells. However, the time and triggering of differentiation remain to be defined (Pujol et al. 1997); this may be easier to do in marsupials.

In general, clear synaptic differentiations can be observed after 10 days before the onset of function. The inner hair cells mature histologically before the outer hair cells. The first endings to contact IHCs are afferent. Efferents follow, initially contacting the cell body and subsequently retracting to synapse with the afferent dendrites.

Synaptogenesis at the base of OHCs follows a completely different pattern. At about 10 d prior to the onset of hearing, OHCs are densely innervated by afferent endings, as are IHCs. A few days later the synaptic specializations at the base of the OHC, initially abundant, decrease markedly in number, although they are still present at the onset of hearing. However, no efferent endings can be recognized at the latter time.

Efferents may arrive at the OHCs after first influencing the innervation of IHCs. They then make transient axodendritic contacts with the numerous afferents at the base of the OHC and appear to push the afferents away (Pujol 1985). The afferents retract, the large synapses made by efferents upon OHC mature, and the OHC elongates, thus ending a "sensory cell" phase for the OHC.

The regular arrangement of three rows of OHCs and one of IHCs is not achieved at the apex of cochlea in adults; here the hair cells are more disorganized. Furthermore the large efferent synapses made elsewhere on OHCs are rare or absent at the apex, as though the apex of the cochlea had never reached maturity in respect of OHCs and their synapses (Pujol et al. 1997).

Nerve fibers appear in the inner ear at a very early stage of development, and spiral ganglion neuron differentiation occurs well before the hair cells have differentiated into IHC and OHC. The cells migrate mainly from the acoustic placode rather than from the neural crest (Deol 1967). Although nerve fibers are early to appear, their myelination is among the last stages of neural development in the cochlea. In kittens, myelination takes up to 6 months to complete (Romand and Romand 1982).

8.4.3 Ultrastructural Development of the Cochlea in *Monodelphis*

At 6 days, OHCs and IHCs are not morphologically distinguishable, although cells corresponding to these are in position. All cells, including hair cell precursors, are covered with microvilli. Neurites, swollen by excitotoxicity, are adjacent to the bases of these cells (Fig. 8.5A). Thus nerve fibers appear to have arrived *prior* to HC differentiation. At 13 days, the hair cells have differentiated, stereocilia are present at their tops and nerve fibers at their bases (Fig. 8.5B); electron-dense contacts suggest the beginning of synapses. The lateral cysternae of OHC are forming but are incomplete. No cysternal pillars are present. The nerve fibers beneath the habenula perforata are unmyelinated.

The tectorial membrane appears at 13 days and by 17 days it shows a sequence of maturity from base to apex. Hair cells are visible through the membrane at the apex but not at the base. By 26 days the cochlea is that of a hearing animal. The cuticular plate is adult-like, as are the number of mitochondria, the presence of lateral efferents to the OHCs, and, within the latter, cisternal pillars. Myelinated nerve fibers are present below the habenula perforata, but spiral ganglion cells are not yet myelinated. The picture is very similar at 30 d, confirming anatomically what has been shown electrophysiologically (Reimer 1996): that *Monodelphis* now hears.

8.5 Anatomical Changes in the Brain

The major division into fore-, mid- and hindbrain is quite distinct when the young marsupial appears in the pouch. The brains of younger pouch-young

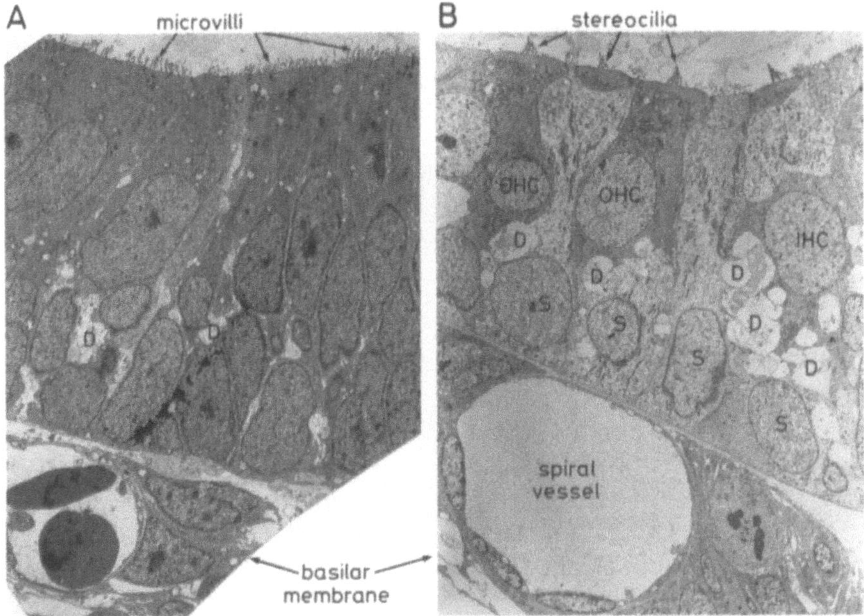

Fig. 8.5A,B. Transmission electron micrographs of the developing organ of Corti in 6-day-old (A) and 13-day-old (B) *Monodelphis* opossums. *IHC, OHC* Nuclei of inner and outer hair cells; *S* nuclei of supporting cells; *D* dendrite of primary afferents swollen by excitotoxicity. Original magnifications: **A** 1500×; **B** 1000×. (L. Aitkin, R. Pujol, and G. Humbert, unpubl. observ.)

quolls are extremely soft, virtually gel-like, because they lack myelin and some other structural material in the extracellular matrix. The mesencephalon dominates the brain during about the first third of pouch life (Fig. 8.6). In quolls the separate bumps of the superior and inferior colliculi (the latter marked with a spot of DiI tracer, see Sect. 8.5.3) are easily differentiated from about 25 days onward (Fig. 8.6). The neocortex grows much more rapidly than the mesencephalon: the width of the midbrain at the collicular level approximately doubles from 29–86 days, whereas the cerebral hemispheres at 86 days are at least four times the widths of those at 29 days. As the cerebral cortex grows, it extends caudally as well as rostrally and laterally, and gradually covers the rostral midbrain. At the same time the cerebellum extends rostrally and hides the caudal part of the inferior colliculus. At maturity, both colliculi are visible between the cerebrum and cerebellum, a feature of marsupials.

Nelson (1988) has reviewed the literature concerning the developing nervous systems of marsupials. He points out that the only system likely to be functional early in pouch life will be the trigeminal somatosensory system, which is able to provide most of the information needed to regulate motor responses in the pouch. Olfactory cues will become increasingly important during pouch life, and there is a corresponding growth in the olfactory bulbs and higher olfactory centers during this period. Nelson emphasizes the need to develop a staging system based on the relative degree of development of a large number of external and

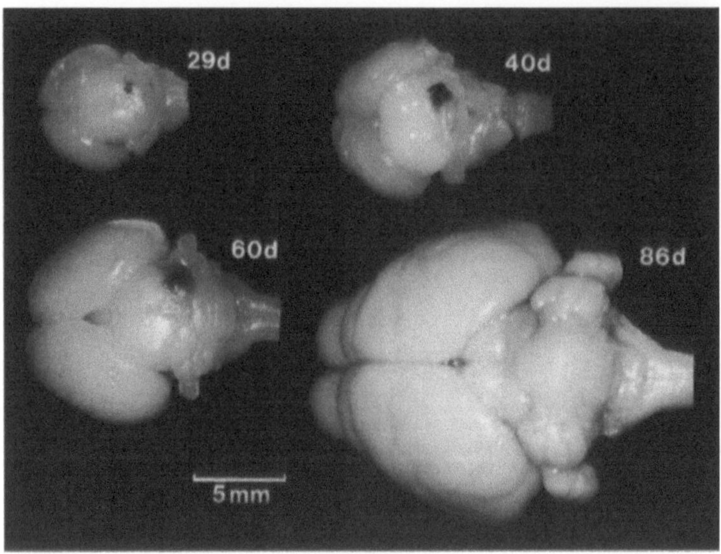

Fig. 8.6. Dorsal views of the brains of four pouch-young quolls at the stated ages. The *dark spots* (injections of the tracer DiI) on the mesencephalon of the three younger animals indicate the locations of the inferior colliculus. Note how the proportion of brain volume occupied by the mesencephalon decreases with age as the cerebral cortex grows. (Aitkin et al. 1994a, with permission of Wiley-Liss, Inc.)

internal characteristics of embryos, in order to allow comparsion with better-known eutherian species.

8.5.1 Cytogenesis, Migration and Formation of Nuclei

Cytogenesis is studied by injecting tritiated thymidine into the developing neonate (Angevine and Sidman 1961). Thymidine is taken up into the deoxyribonucleic acid of dividing cells, but is cleared from the circulation well within the day of injection. As a consequence, cells that contain label in the adult must have been undergoing division at the time of injection, establishing birth dates to within a day. Neurogenesis in the auditory system has been studied in a number of eutherians (Taber-Pierce 1973; Altman and Bayer 1981; Cooper and Rakic 1981) and two marsupial species, quolls (Aitkin et al. 1991) and brushtail possums (Sanderson and Aitkin 1990).

In the polyprotodont quoll, neurons in the ventral cochlear nucleus are generated prior to 3 days pouch life, in the superior olive at 5–7 days, and in the dorsal cochlear nucleus over a prolonged period. Inferior collicular neurogenesis lags behind that in the medial geniculate, the latter taking place between days 3–9, and the former between days 7–22. Neurogenesis begins in the auditory cortex on day 9 and is completed by about day 42.

A very similar time course occurs in the diprotodont brushtail possum. In the cochlear nucleus large neurons are born by postnatal day 5 and small cells (including granule cells) over a later, protracted period. Neurogenesis in the IC occurs between days 5–28 and that in the MG from about day 7–12. Thus, in both marsupial species, neurogenesis is complete in the medullary auditory nuclei before that in the midbrain commences and is concluded in the MG before the IC. Neurogenesis in the auditory cortex occurs over a prolonged period (Fig. 8.7). The first cells to be born (days 5–12) are destined for the deepest layers; as development proceeds the newborn cells are found in more and more superficial layers until about day 39, when most neurogenesis is completed. This inside-out pattern of cortical development, first documented by Angevine and Sidman (1961), also occurs in the quoll.

The studies of Sanderson and Aitkin (1990) and Sanderson and Weller (1990) allow comparison between neurogenesis in the auditory, visual, and sensorimotor systems in the one species, the brushtail possum. Neurons in the subcortical visual pathway are generated between postnatal days 5 and 21 (as for the IC and MG), whereas neurogenesis in the visual cortex continues until about day 68, later than the auditory cortex (in keeping with the later eye-opening versus onset of

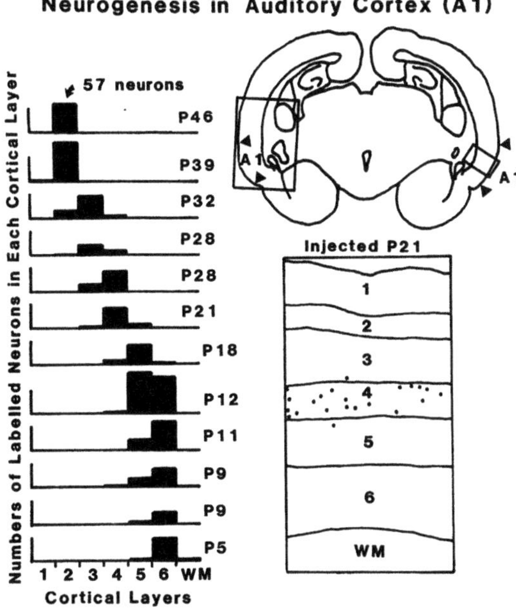

Fig. 8.7. Neurogenesis in the auditory cortex of the brushtail possum. The line drawing (*top right*) shows the location of primary auditory cortex (*AI*, demarcated by *arrowheads*). The series of histograms on the *left* shows the laminar distribution of neurons, labeled by a series of tritiated thymidine injections, generated between postnatal (*P*) day *P5* and *P46*. Each individual graph was generated by counting the number of labeled neurons in each layer of a section through the auditory cortex. The plot at *lower right* shows the distribution of labeled neurons in a narrow strip (*small rectangle on the right side of the brain section*) of the auditory cortex of a possum injected with tritiated thymidine on *P21*. The cortical layers are *numbered*; *WM* white matter. (Sanderson and Aitkin 1990, with permission)

hearing). Neurogenesis in thalamic sensorimotor nuclei begins early (around the time of birth) and is concluded by day 11. Formation of neurons in the sensorimotor cortex and basal ganglia occurs during the first two months of postnatal life and in the cerebellum during the first 3 months. In general, neurogenesis begins earlier than in eutherian mammals but is more protracted; this protracted development offers great advantages for developmental studies (Sanderson and Weller 1990).

Once cells are born they migrate to their ultimate locus and form connections. During this time, the nuclei of the auditory pathway are forming. The inferior colliculus of the quoll is first recognizable cytoarchitecturally in pouch young aged 23 days, when it is bordered by a cell-sparse ring of tissue (Aitkin et al. 1994a). By this time, the labeling patterns following injections of tritiated thymidine made on days 7–9 suggest that migration of cells to the inferior colliculus from the ventricular germinal zone has been largely completed. The period 45–50 days is associated with a large expansion of cell volume and a concomitant decrease in packing density. At 81 days, close to the time when the young move out of the pouch, the adult cytoarchitecture – a central nucleus flanked by dorsal and lateral cortical regions – is clear.

8.5.2 Development of Dendrites of the Inferior Colliculus

The Golgi technique impregnates all processes in a small proportion of cells, and so is a useful qualitative indicator of dendritic growth. When applied to quoll pouch young of 19 days, short stubby processes are impregnated extending for up to 20 µm from irregular cell bodies 5–7 µm in the longest diameter (Fig. 8.8). Small (less than 1 µm) spherical varicosities occur along the processes (small arrows, 19 d), sometimes aggregating into clumps (large arrowhead, 19 d). At this age, processes cannot be confirmed as either axons or dendrites; they rarely branch beyond the cell body, and they lack spines (inset, 19 d).

By day 31 many unequivocally neuronal profiles are impregnated (Figs. 8.8, 8.9). Processes are much more numerous than on day 19 and extend considerable distances without many branches (e.g., neuron to left of inset, Fig. 8.8, 31 d). For others, processes ramify profusely in the vicinity of the cell body (e.g., neuron above inset, 31 d). In general, few secondary and no tertiary branches are observed. Some processes are spiny (e.g., inset, 31 d) and others smooth. Thus, some processes are certainly young dendrites, but no axons are unequivocally identified. At 31 days a complex network of radial glia, arranged around the aqueduct, bear somata, probably those of migrating neurons, along their lengths (Fig. 8.9, 31 d, right).

At 45 days individual neurons (e.g., Fig. 8.8, 45 d) are indistinguishable from adult neurons in appearance and complexity of branching. Dendrites are usually spiny (e.g., inset, 45 d) and exhibit secondary, tertiary, and occasionally quaternary branches. Presumed axons make shallow inflexions or bends shortly after leaving the cell body and are smooth, lacking varicosities. Radial glia are still present near the third ventricle and aqueduct.

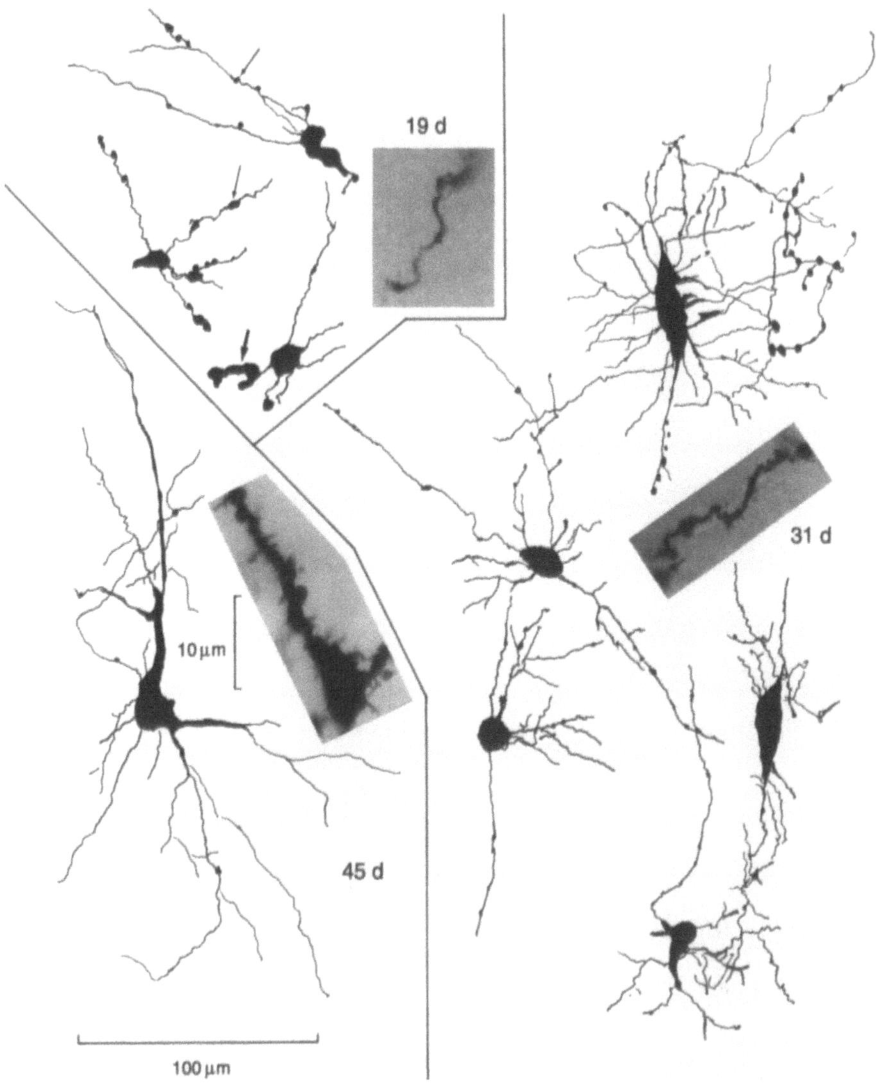

Fig. 8.8. Drawings (lower calibration 100 µm) made with a Leitz Micropromar projection microscope of cells stained with the Ramon-Moliner modification of the Golgi-Cox technique, located in the vicinity of the inferior colliculus of Northern quolls at 19 days (*19 d*) or in its central nucleus at *31* and *45 days*. Photomicrograph *insets* (calibration 10 µm) show portions of dendritic trees from cells in each age group. Spherical varicosities (*small arrows*) which sometimes clumped into large profiles (*large arrow*)

The axons now present in the IC impose a stratified appearance on the nucleus (Fig. 8.9, 45 d). Fibers sweep around the nucleus, creating a cell-sparse zone. This has been observed in Nissl material from pouch young aged 36–50 days (Aitkin et al. 1994a). Within the IC the feltwork of processes varies in density, adding to the sense of orientation imparted by the external fiber capsule (Fig. 8.9, 45 d).

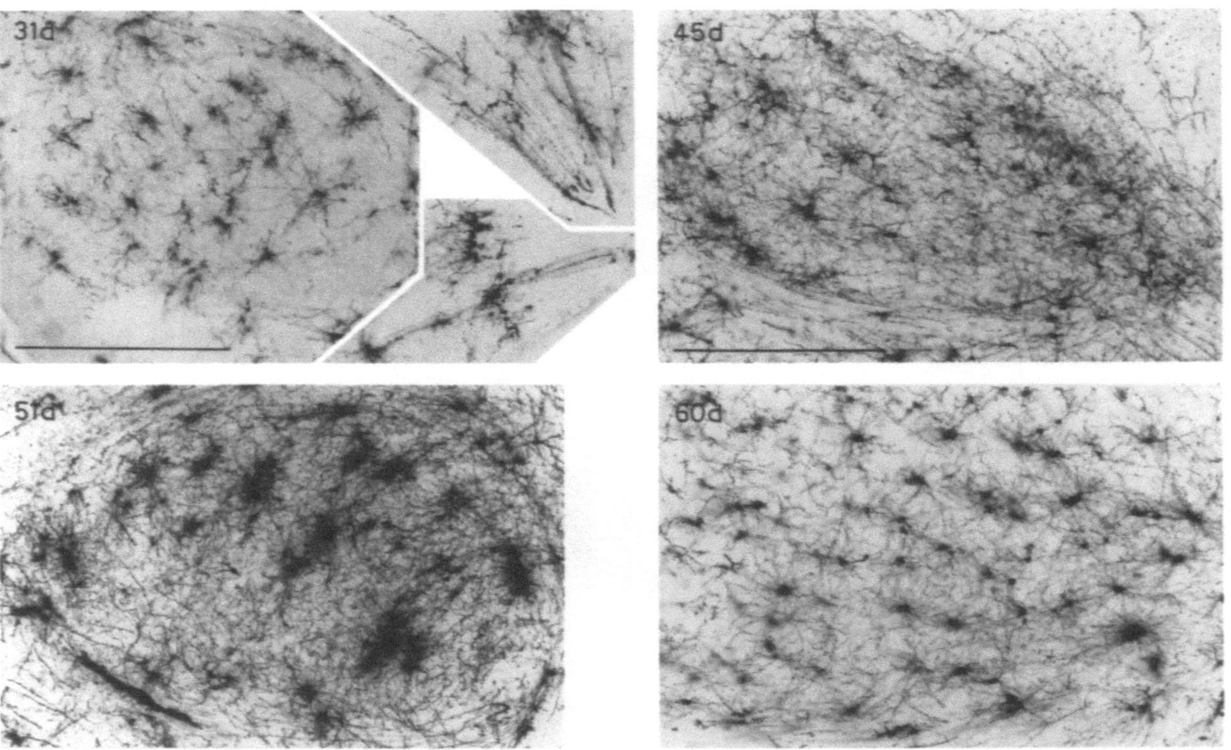

Fig. 8.9. Photomicrographs of the inferior colliculus of quolls: *frontal sections*, stained by the Ramon-Moliner modification of the Golgi Cox technique at *31, 45, 51 and 60* days pouch life. Calibration 500 µm. Dorsal up, section from 51 days, medial to left; for all others medial to right

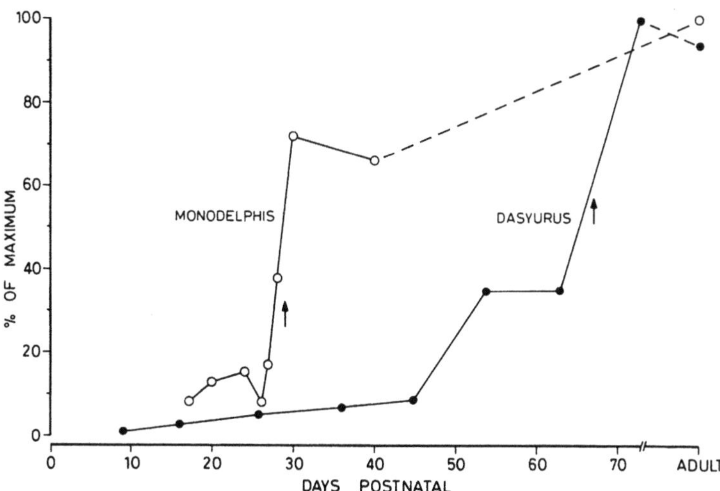

Fig. 8.10. Synapses were counted in a specified area ($97\,\mu m^2$) from the inferior colliculus of quolls (*Dasyurus*) and *Monodelphis* opossums and compared with the numbers of cells in the same area. Each point is the average of at least three sections. The numbers were normalized to the percentage of the maximum number of synapses/cells in each species (*ordinate*) and plotted against postnatal age in days (*abscissa*). Note the steep increase (*arrows*) between 26 and 30 days for *Monodelphis* and between 63 and 73 days in *Dasyurus*, at times when hearing is just beginning

Cellular profiles with the characteristics of glial cells are not apparent at 19, 31 or 45 days, but bushy processes – presumably fibrous astrocytes – become visible later in the IC (arrows, 51 d, 60 d, Fig. 8.9). It has been suggested, based on Nissl material, that a proliferation of glial cells occurs in the IC after 50 days pouch life in the quoll (Aitkin et al. 1994a). The IC at 60 days, just prior to the onset of hearing, is dominated by the widely ramifying dendrites of its constituent neurons (Fig. 8.9, 60 d). These are arranged in fairly regular rows along the trajectories of the fibers of the IC, imparting a layered appearance to the structure. Such a laminated IC in the adult has been discussed previously as being a characteristic of both eutherians and, probably, marsupials (Chap. 6).

8.5.3 Development of Connections and Synapses in the Inferior Colliculus

Injections of horseradish peroxidase (HRP) into the IC of neonatal Virginia opossums at postnatal day 5 label neurons in a band adjacent to, but not within, the presumptive cochlear nucleus (Willard and Martin 1986). By day 15, similar injections label cells within the nucleus proper, and by day 22 (about 37 days after conception) these cells can be classified as large neurons of DCN and VCN. The authors conclude that the axons of these cells reach the inferior colliculus while migrating from the cytogenetic zone to the cochlear nucleus.

Connections appear to be established much later in Northern quolls, although it is clear from the cytogenetic work cited above that cells destined for the cochlear nucleus are born prior to or soon after birth and reach their destination early. Transport of the tracer substance DiI, after deposition into the IC of pouch-young quolls, reveals the presence of connections from the medullary auditory nuclei on postnatal day 36, but the results are equivocal on day 26 (Aitkin et al. 1996b). This timeline for the establishment of connections is in reasonable concordance with the observations that migration of many of the neurons of the IC has occurred by day 22, when the central nucleus first becomes recognizable as a cytoarchitectural entity, and is complete by day 29 (Aitkin et al. 1994a). The targets for some of these axons, dendritic spines, proliferate between days 31–45 (Fig. 8.8). Myelin first appears in the IC of quolls at day 73. At this time the myelin sheaths are composed of approximately ten layers (Aitkin et al. 1996b). Given the dense and widespread myelination of fibers of the IC of adults (e.g., Rockel and Jones 1973b), it is likely that in quolls a major increase in myelination occurs between the end of pouch life and adulthood.

Synapses are present in the IC of pouch-young quolls at day 9, but they are very rare (Aitkin et al. 1996b). Their numbers increase throughout pouch life, reaching a peak at 73 days and falling to adult values thereafter. A comparison of synaptic and neuronal packing densities suggests that, after a steady increase in synapses/ neuron between 9–63 days, synaptic density rises steeply at about the time hearing begins, to reach adult values at day 73 (Fig. 8.10). These results were obtained with animals from the same litter, giving homogeneity to the genetic background of the samples. Synaptogenesis in the inferior colliculus of *Monodelphis* increases between days 26–30, also the time when hearing begins in this species (Fig. 8.10). There is a further, slower increase to the adult value from day 40 in *Monodelphis* (Fig. 8.10).

Ultrastructural studies of the inferior colliculus have been carried out in rats (Pysh 1969, 1970; Ribak and Roberts 1986) and in cats (Jones and Rockel 1973; Rockel and Jones 1973b; Oliver 1985), but none have examined the question of synaptogenesis in relation to hearing onset. Reynolds (1975) observed synaptic contacts in the inferior colliculus of kittens aged 3 days, at about the time hearing begins in this species (Foss and Flottorp 1974). The synapses identified in quolls in the first 3 weeks of pouch life are unlikely to have been provided by ascending afferent terminals, since these have not reached the IC until some time after day 26. They presumably represent the points of contact between developing neurites or fiber systems within the IC.

8.5.4 Concerning the Determinants of Synaptogenesis

What is the function of synaptic contacts at an age when synaptic transmission between neurons may not have begun? Others have reported synapses at equivalent early ages – for example, in the temporal cortex of rats at 16d gestation (König et al. 1975) and in the human spinal cord of 10-mm human embryos (4–5 weeks of gestation; Okado 1981). Okado suggested that these "synapse-like contacts" might be axoglial, and it is conceivable that the first "synapselike contacts" are the result of genetic programs providing mechanisms for growing nerve fibers to make chemical contacts with other neural elements. These may have no relation to sensory input; the latter may become more and more important in steering synaptogenesis and take over later in development.

Synaptic counts have been made in the visual cortex of cats (gestation period 66 days), from 37 days gestation to adulthood (Cragg 1975). A few synapses are present as early as 3 weeks before birth, but most synaptogenesis occurs 8–37 days after birth. Synapse development in the visual cortex of cats thus begins before light-evoked impulses can enter the visual system, but most synaptogenesis occurs immediately after the eyes open. Interference with the opening of the eyes in newborn cats alters the activity-dependent synaptic competition between geniculocortical axon terminals in the visual cortex (Shatz 1990).

The rapid increase in synaptic density in the IC of pouch-young quolls and *Monodelphis* opossums occurs at about the time the ear canal becomes patent and hearing has begun (see above). This suggests that environmental sound may be an important stimulus for synaptogenesis. It would be expected that patterns of neuronal activity, such as those generated by environmental sound, will be vital in refining connections guided initially by genetically determined molecular mechanisms (for review, see Goodman and Shatz 1993). Genetic programs may require a normally functioning peripheral auditory system to exist before permitting central synaptogenesis to proceed rapidly.

It is likely that the major connections of the auditory system are fully formed before hearing begins. Impulses can be evoked in the auditory cortex of kittens by electrical stimulation of the cochlear nerve a few days prior to the onset of hearing (Marty and Thomas 1963), so the auditory pathway of this species must be capable of transmitting information but does not do so due to a lack of input caused by the high attenuation to sound of the immature auditory periphery. In tammar wallabies, neurons recorded in the auditory nerve and CNC discharge spontaneously in rhythmic bursts some days before hearing begins (Gummer and Mark 1994). Similar spontaneous activity is observed in other immature sensory systems and may have a role in stimulating synaptogenesis, perhaps accounting for the accelerated synaptogenesis which occurs in quolls around the time hearing begins.

The auditory environment of the pouch-young marsupial is quite different to that of the eutherian fetus, where the attenuation provided by the amnion and the mother's body will filter out higher frequencies of external origin and enhance the endogenous low frequencies of the heart beat and other sounds of thoracic origin. The developing marsupial does not receive this barrage of low-frequency sound but is subject to the whole gamut of broad-spectrum environmental sound. Nonetheless, for both marsupial and eutherian, the last developmental stage prior to the onset of hearing is likely to be the pneumatization of the middle ear and opening of the external ear canal (Pujol and Hilding 1973).

8.6 Future Research Directions

The development of the auditory system of marsupials has only been touched upon at the time of writing. Some beginnings have been made by longitudinal study of the IC in one species, the Northern quoll, and some supporting material is available from brushtail possums and the *Monodelphis* opossum. A number of key questions about the development of the auditory system are readily amenable to investigation in marsupials. When do inner and outer hair cells become distinguishable and what are the determinants of their differentiation? When do cochlear nerve fibers innervate the hair cells, and what differences in time courses and patterns of innervation occur with afferent and efferent fibers?

Determinants of growth and maturation can be examined readily with marsupials. The accessibility of the developing marsupial provides opportunities for altering aspects of the external environment and for mechanical and surgical interference with the developing young. For example, Nelson (1988) mentions the possibilities of tissue cultures from early pouch young and of dietary alterations (since the "embryo" receives its nutrition from milk via a teat), to which the developing nervous system is especially sensitive.

Questions of plastic change in the auditory nervous system (e.g., Robertson and Irvine 1989; Willott et al. 1993) and critical periods during development (e.g., Hubel and Wiesel 1970; Moore and Irvine 1981) can be addressed by experimentation on developing marsupials. Can experimental manipulations alter the topographic organization of the auditory system and, if so, at what period during development? The importance of the state of development of each level of the auditory system (cochlea, medulla, midbrain, etc.) on the growth of successive levels is worthy of study, given the intriguing nonsequential neurogenesis of the auditory pathway. The removal of the developing otocyst or lesions to the developing midbrain can be followed up by examination of the alterations in connections of the auditory pathway. The results of such investigations are likely to have importance well beyond the question of marsupial development per se.

Chapter 9

Conclusions

Distinctive features have been demonstrated in marsupials at all levels of the auditory system and in hearing ranges and vocal behaviors. Briefly, audiograms of marsupials indicate a range of sensitivities comparable to eutherians, but they appear generally less sensitive, either in minimum threshold or in range of frequencies. This could be due to the weight and arrangements of middle-ear structures, but there are other possibilities. As a whole, adult marsupials are not a very vocal order, although some arboreal genera are the exceptions to this rule. The neonates of some species are very vocal, and spectra of their vocalizations are closely correlated to the best frequency of hearing of the adult. There are great similarities between marsupials and eutherians in the organization of the brain auditory pathway, but certain differences do stand out, such as the arrangements of the cochlear nuclei, the organization and projections of the superior olive, the sensitivity of neurons to broadband acoustic stimuli, and the parcellation of auditory cortical fields and their interhemispheric projections. Only further research can tell if these features of neural organization underlie subtly different outcomes in hearing and vocalization.

9.1 Hearing Sensitivity

Insensitivity to airborne sound has been observed in studies of three opossum species (Ravizza et al. 1969; Frost and Masterton 1994). In studies in which ABR audiograms have been measured, poor thresholds have been noted in the feathertail glider (Chap. 4) and quite variable thresholds in tammar wallabies (Cone-Wesson et al. 1997). However, this feature does not seem to be a marsupial trait, because poor thresholds have been observed in eutherian species, such as hedgehogs (Ravizza et al. 1969; Batzri-Israeli et al. 1990) and mole rats (Bronchti et al. 1989), and at least some marsupials have very sensitive hearing (e.g., brushtail possum and quoll; see Chap. 3). One of the opossum species (*Monodelphis*), when tested using a different behavioral paradigm, may be more sensitive than first thought (Reimer and Baumann 1995).

An alternative interpretation is that insensitive hearing among mammals is an indicator of the primitiveness of a species within an order. The two surviving monotremes, platypus and echidna, considered by many to be among the most primitive mammals, appear also to be insensitive to sound (Aitkin and Johnstone 1972; Gates et al. 1974). In a provocative essay, Masterton and his colleagues (1969) analyzed different parameters of hearing across groups of animals to approximate a phylogenetic sequence of man's ancestors. They argued that high-frequency hearing is a characteristic unique to mammals but that low-frequency hearing improved in mankind's line of descent. With regard to hearing sensitivity, they concluded that "the only large difference in general sensitivity among mammals in the phyletic sequence that may be both real and significant occurs between levels at the lowest stages (i.e. opossum-hedgehog-treeshrew levels, phyletically . . ." (p. 979). It is likely that modern marsupials, particularly those in Australia, have evolved their hearing to suit their ecology. Some have high sensitivity (e.g., brushtail possums), whereas others may be insensitive (e.g., feathertail gliders).

9.2 Relationship Between Hearing Range and Vocal Behavior

Marsupials as a group do not make many sounds audible to humans, but there are exceptions among arboreal species. The loud and complex calls of yellow-bellied and sugar gliders, and ringtail and striped possums have been noted – the former three species are highly social, yet macropodids, also social, are not very vocal. The only loud calls of nocturnal carnivores seem to be made during aggressive interactions.

The elaboration of vocal behavior in a given mammalian species is likely to be the product of the interaction of a number of factors. There is no doubt that the species with the most elaborate vocalizations, humans, has unique neural machinery to make this possible. Species that are at a low level of evolution within an order ("primitive") seem to be more mute than those at a higher level (e.g., polyprotodont vs diprotodont marsupials; prosimians compared with humans; monotremes compared with eutherians). Species that are social may make more use of vocal communication than solitary animals (e.g., ringtail possums compared with greater gliders; guinea pigs vs. moles; vervet monkeys compared with gorillas). Finally, a creature's ecology may highlight features of vocal behavior or render it unimportant (arboreal species with restricted visual cues; nocturnal carnivores which use their hearing for prey detection, when conspecific calling would be a nuisance).

When viewed in terms of these factors, adult marsupials do not appear to stand out as particularly unusual. However, the calls made by deaf and blind neonatal marsupials ("pouch-young") are interesting, because the spectra of these calls, in the two species studied, are closely attuned to the best frequency of hearing of the adult. This may be a survival strategy for marsupials and not found in eutherians, but further research is needed to test this hypothesis.

9.3 The Marsupial Auditory Pathway

The available neuroanatomic and neurophysiological information indicates that most features of the organization of the auditory pathway of marsupials have parallels with eutherians. The outcomes are likely to be the same, but the means to the end may differ. It is beyond the scope of this volume to speculate on the evolutionary events that may have provided the different ways in which, for example, the cochlear nuclei are arranged in the different mammalian orders. However, it must be pertinent that similar organizational features have been observed in "primitive" marsupials, such as the didelphids, in less primitive diprotodonts, and in marsupials of differing habitats and ecologies. Judging from the restriction of diprotodontids to Australasia, the time at which the divergence of the marsupial and eutherian auditory pathways occurred must have been well before the continental changes leading to the break up of Gondwanaland.

Given the essential similarities of the auditory pathways of marsupials and eutherians, yet considering the immaturity at birth of the marsupial, the organization of the auditory pathway of marsupials may well prove extremely important to the developmental biologist. Factors determining the wiring of the auditory pathway can be dissected in neonatal marsupials and their normal influence prevented or modified. The outcomes of such studies will have significance not only to auditory neuroscientists but to neurobiologists in general.

References

Abbie AA (1937) Some observations on the major subdivisions of the Marsupialia with special reference to the position of the Peramelidae and Caenolestidae. J Anat 71: 429–436

Abbie AA (1939) The origin of the corpus callosum and the fate of structures related to it. J Comp Neurol 70: 9–44

Abbie AA (1940) Cortical lamination in the Monotremata. J Comp Neurol 72: 429–467

Adams JC (1979) Ascending projections to the inferior colliculus. J Comp Neurol 183: 519–538

Aertsen AMHJ, Smolders JWT, Johannesma PIM (1979) Neural representation of the acoustic biotope: on the existence of stimulus-event relations for sensory neurons. Biol Cybern 32: 175–185

Aitkin L (1986) The auditory midbrain. Humana Press, Clifton, New Jersey

Aitkin L (1990) The auditory cortex. Chapman and Hall, London

Aitkin L (1995) The auditory neurobiology of marsupials. A review. Hear Res 82: 257–266

Aitkin L (1996) The anatomy of the cochlear nuclei and superior olivary complex of arboreal Australian marsupials. Brain Behav Evol 48: 103–114

Aitkin LM, Gates GR (1983) Connections of the auditory cortex of the brush-tailed possum, *Trichosurus vulpecula*. Brain Behav Evol 22: 75–88

Aitkin LM, Johnstone BM (1972) Middle-ear function in a monotreme: the echidna (*Tachyglossus aculeatus*). J Exp Zool 180: 245–250

Aitkin L, Jones R (1992) Azimuthal processing in the posterior auditory thalamus of cats. Neurosci Lett 142: 81–84

Aitkin LM, Kenyon CE (1981) The auditory brain stem of a marsupial. Brain Behav Evol 19: 126–143

Aitkin LM, Moore DR (1975) Inferior colliculus II. Development of tuning characteristics and tonotopic organization in the central nucleus of the neonatal cat. J Neurophysiol 38: 1208–1216

Aitkin LM, Nelson JE (1989) Peripheral and central auditory specialization in a gliding marsupial, the feathertail glider (*Acrobates pygmaeus*). Brain Behav Evol 33: 325–333

Aitkin L, Park V (1993) Audition and the auditory pathway of a vocal new world primate, the common marmoset. Prog Neurobiol 41: 345–367

Aitkin LM, Webster WR (1972) Medial geniculate body of the cat: organization and responses to tonal stimuli in the ventral division. J Neurophysiol 35: 365–380

Aitkin LM, Bush BMH, Gates GR (1978a) The auditory midbrain of a marsupial: the brushtailed possum (*Trichosurus vulpecula*). Brain Res 150: 29–44

Aitkin LM, Dickhaus H, Schult W, Zimmermann M (1978b) External nucleus of the inferior colliculus: auditory and spinal somatosensory afferents and their interactions. J Neurophysiol 41: 837–847

Aitkin LM, Gates GR, Kenyon CE (1979) Some peripheral auditory characteristics of the marsupial brush-tailed possum, *Trichosurus vulpecula*. J Exp Zool 209:317–322

Aitkin LM, Gates GR, Phillips SC (1984) Responses of neurons in inferior colliculus to variation in sound-source azimuth. J Neurophysiol 52: 1–17

Aitkin LM, Byers M, Nelson JE (1986a) Brainstem auditory nuclei and their connections in a carnivorous marsupial, the Northern native cat (*Dasyurus hallucatus*). Brain Behav Evol 29: 1–16

Aitkin LM, Irvine DRF, Nelson JE, Merzenich MM, Clarey JC (1986b) Frequency representation in the auditory midbrain and forebrain of a marsupial, the Northern native cat (*Dasyurus hallucatus*). Brain Behav Evol 29: 17–28

Aitkin L, Nelson J, Farrington M, Swann S (1991) Neurogenesis in the brain auditory pathway of a marsupial, the Northern native cat (*Dasyurus hallucatus*). J Comp Neurol 309: 250–260

Aitkin L, Nelson J, Farrington M, Swann S (1994a) The morphological development of the inferior colliculus in a marsupial, the Northern quoll (*Dasyurus hallucatus*). J Comp Neurol 343: 532–541

Aitkin LM, Nelson JE, Shepherd RK (1994b) Hearing, vocalization and the external ear of a marsupial, the Northern quoll, *Dasyurus hallucatus*. J Comp Neurol 349: 377–388

Aitkin LM, Nelson JE, Shepherd RK (1996a) Development of hearing and vocalization in a marsupial, the Northern quoll, *Dasyurus hallucatus*. J Exp Zool 276: 394–402

Aitkin LM, Nelson JE, Martsi-McClintock A, Swann S (1996b) Features of the structural development of the inferior colliculus in relation to the onset of hearing in a marsupial, the Northern quoll, *Dasyurus hallucatus*. J Comp Neurol 375: 77–88

Alford BR, Ruben RJ (1963) Physiological, behavioral and anatomical correlates of the development of hearing in the mouse. Ann Otol Rhinol Laryngol 72: 237–247

Altman J, Bayer SA (1981) Time of origin of neurons of the rat inferior colliculus and the relations bettween cytogenesis and tonotopic order in the auditory pathway. Exp Brain Res 42: 411–423

Angevine JB, Sidman RL (1961) Autoradiographic study of cell migration during histogenesis of cerebral cortex in the mouse. Nature 192: 766–788

Aplin KP, Archer M (1987) Recent advances in marsupial systematics with a new syncretic classification. In: Archer M (ed) Possums and opossums: studies in evolution. Surrey Beatty and Sons, Sydney, pp xv–lxxii

Archer M (1984) Origins and early radiations of marsupials. In: Archer M, Clayton G (eds) Vertebrate zoogeography and evolution in Australasia. Hesperian Press, Sydney, pp 585–625

Archer M, Hand SJ, Godthelp H (1991) Riversleigh. Reed, Sydney

Batzri-Izraeli R, Kelly JB, Glendenning KK, Masterton RB, Wollberg Z (1990) Auditory cortex of the long-eared hedgehog (*Hemiechinus auritus*). I. Boundaries and frequency representation. Brain Behav Evol 36: 237–248

Biggins JG (1984) Communication in possums: a review. In: Smith AP, Hume ID (eds) Possums and gliders. Australian Mammal Society, Sydney, pp 35–57

Birnholz JC, Benacerraf BR (1983) The development of human fetal hearing. Science 222: 516–518

Bobbin RP, May JG, Lemoine RL (1979) Effect of pentobarbital and ketamine on brain stem auditory potentials. Arch Otolaryngol 105: 467–470

Borg E (1973) On the neuronal organization of the middle ear reflex. A physiological and anatomical study. Brain Res 49: 101–123

Bronchti G, Heil P, Scheich H, Wollberg Z (1989) Auditory pathway and auditory activation of primary visual targets in the blind mole rat (*Spalax ehrenbergi*). I. 2-Deoxyglucose study of subcortical centers. J Comp Neurol 284: 253–274

Brown CH, Waser PM (1988) Environmental influences on the structure of primate vocalizations. In: Todt D, Goedeking P, Symmes D (eds) Primate vocal communication. Springer Berlin Heidelberg New York, pp 51–66

Brownell WE, Bader CR, Bertrand D, deRibaupierre Y (1985) Evoked mechanical responses in isolated outer hair cells. Science 227: 194–196

Calford MB (1983) The parcellation of the medial geniculate body of the cat defined by the auditory response properties of single units. J Neurosci 3: 2350–2364

Calford MB, Aitkin LM (1983) Ascending projections to the medial geniculate body of the cat: evidence for multiple, parallel auditory pathways through thalamus. J Neurosci 3: 2365–2380

Calford MB, Webster WR (1981) Auditory representation within principal division of cat medial geniculate body: an electrophysiological study. J Neurophysiol 45: 1013–1028

Carlier E, Lenoir M, Pujol R (1979) Development of cochlear frequency selectivity tested by compound action potential tuning curves. Hear Res 1: 197–201

Cheney D, Seyfarth R (1982) How vervet monkeys perceive their grunts. Anim Behav 30: 739–751

Clemens WA, Richardson BJ, Baverstock PR (1989) Biogeography and phylogeny of the metatheria. In: Walton DW, Richardson BJ (eds) Fauna of Australia. Mammalia vol 1 B. Australian Government Publishing Service, Canberra, pp 527–548

Coles RB, Guppy A (1986) Biophysical aspects of directional hearing in the tammar wallaby, *Macropus eugenii*. J Exp Biol 121: 371–394

Coles RB, Hill KG (1981) Ear movements in the tammar wallaby *Macropus eugenii*. Proc Aust Physiol Pharmacol Soc 12: 169P

Cone-Wesson BK, Hill KG, Liu G-B (1997) Auditory brainstem response in tammar wallaby (*Macropus eugenii*). Hear Res 105: 119–129

Cook CJ, Williams C, Gluckman PD (1987) Brainstem auditory evoked potentials in the fetal sheep, in utero. J Dev Physiol 9: 429–439

Cooper ML, Rakic P (1981) Neurogenetic gradients in the superior and inferior colliculi of the Rhesus monkey. J Comp Neurol 202: 309–334

Cowley AR (1973) The nuclei of the cochlear nerve of the red kangaroo, *Megaleia rufa*. J Hirnforsch 14: 287–301

Cragg BG (1975) The development of synapses in the visual system. J Comp Neurol 160: 147–166

Crewther DP, Crewther SG, Sanderson KG (1984) Primary visual cortex in the brushtailed possum: receptive field properties and cortico-cortical connections. Brain Behav Evol 24: 184–197

Croft DB (1982) Communication in the Dasyuridae (Marsupialia): a review. In: Archer M (ed) Carnivorous marsupials. Royal Zoological Society of New South Wales, Sydney, pp 291–309

Dallos P (1984) Peripheral mechanisms of hearing. In: Darian-Smith I (ed) Handbook of physiology, sect 1. The nervous system, vol 8. Sensory processes. American Physiological Society, Bethesda, pp 595–637

Dallos P, Harris D, Özdamar O, Ryan A (1978) Behavioral, compound action potential and single unit thresholds: relationships in normal and abnormal ears. J Acoust Soc Am 64: 151–157

Davis H, Hirsh SK, Turpin LL, Peacock ME (1985) Threshold sensitivity and frequency specificity in auditory brainstem response audiometry. Audiology 24: 54–70

Davis M, Gendelman DS, Tischler MD, Gendelman PM (1982) A primary acoustic startle circuit – lesion and stimulation studies. J Neurosci 2: 791–805

Dempster E (1994) Vocalisations of adult Northern quolls, *Dasyurus hallucatus*. Aust Mammal 17: 43–49

Deol MS (1967) The neural crest and acoustic ganglion. J Embryol Exp Morphol 17: 533–541

Doran AHG (1879) On the morphology of the mammalian ossicula auditus. Trans Linn Soc Lond 1: 371–497

Ebner FF (1969) A comparison of primitive forebrain organization in metatherian and eutherian mammals. Ann NY Acad Sci 167: 241–257

Ehret G (1976) Development of absolute auditory thresholds in the house mouse (*Mus musculus*). J Am Audiol Soc 1: 179–184

Ehret G (1983a) Development of hearing and response behavior to sound stimuli: behavioral studies. In: Romand R (ed) Development of auditory and vestibular systems. Academic Press, New York, pp 211–237

Ehret G (1983b) Psychoacoustics. In: Willott JF (ed) The auditory psychobiology of the mouse. Thomas, Springfield, pp 13–56

Eisenberg JF, Collins LR, Wemmer C (1975) Communication in the Tasmanian devil (*Sarcophilus harrisii*) and a survey of auditory communication in the Marsupialia. Z Tierpsychol 37: 379–399

Epple G (1968) Comparative studies of vocalization in marmoset monkeys (Hapalidae). Folia Primatol 8: 1–40

Evans EF (1972) The frequency response and other properties of single fibres in the guinea-pig cochlear nerve. J Physiol (Lond) 226: 263–287

Fadem BH, Trupin GL, Maliniak E, VandeBerg JL, Hayssen V (1982) Care and breeding of the gray, short-tailed opossum (*Monodelphis domestica*). Lab Anim Sci 32: 405–409

Farley SD, Lehner PN, Clark T, Trost C (1987) Vocalizations of the Siberian ferret (*Mustela eversmanni*) and comparisons with other mustelids. J Mammal 68: 413–416

Faulstich M, Kössl M, Reimer K (1996) Analysis of non-linear cochlear mechanics in the marsupial *Monodelphis domestica*: ancestral and modern mammalian features. Hear Res 94: 47–53

Fay RR (1988) Hearing in vertebrates: a psychophysics databook. Hill-Fay Associates, Winnetka, Illinois

Faye-Lund H, Osen KK (1985) Anatomy of the inferior colliculus in rat. Anat Embryol 171: 1–20

Fernández C, Schmidt RS (1963) The opossum ear and evolution of the coiled cochlea. J Comp Neurol 121: 151–159

Filan SL (1991) Development of the middle ear region in *Monodelphis domestica*: marsupial solutions to an early birth. J Zool Lond 225: 577–588

Fitzpatrick KA (1975) Cellular architecture and topographic organization of the inferior colliculus of the squirrel monkey. J Comp Neurol 164: 185–208

Fjermedal O, Laukli E (1989) Paediatric auditory brainstem response and pure-tone audiometry: threshold comparisons. Scand Audiol 18: 105–111

Flannery TF (1994) The future eaters. Reed, Sydney

Foss I, Flottorp G (1974) A comparative study of the development of hearing and vision in various species commonly used in experiments. Acta Otolaryngol 77: 202–214

Friedman I, Ballantyne J (1984) Ultrastructural atlas of the inner ear. Butterworths, London

Frost SB (1995) The *Monodelphis* medial geniculate body: comparison of the anatomical connections of the non-cortically and cortically projecting auditory diencephalon in a neurologically primitive mammal. PhD Thesis, Florida State University, Tallahassee

Frost SB, Masterton RB (1992) Origin of auditory cortex. In: Webster DB, Fay RR, Popper AN (eds) The evolutionary biology of hearing. Springer, Berlin Heidelberg New York, pp 655–671

Frost SB, Masterton RB (1993) Thalamocortical-corticothalamic reciprocals and the evolutionary origin of medial geniculate. In: Minciacchi D, Molinari M, Macchi G, Jones EG (eds) Thalamic networks for relay and modulation. Pergamon, Oxford, pp 29–38

Frost SB, Masterton RB (1994) Hearing in primitive mammals: *Monodelphis domestica* and *Marmosa elegans*. Hear Res 76: 67–72

Gaskill SA, Brown AM (1990) The behaviour of the acoustic distortion product, 2fl-f2, from the human ear and its relation to auditory sensitivity. J Acoust Soc Am 88: 821–839

Gates GR, Aitkin LM (1982) Auditory cortex in the marsupial possum *Trichosurus vulpecula*. Hear Res 7: 1–11

Gates GR, Aitkin LM (1984) The auditory system of the common brushtail possum – from cochlea to cortex. In: Smith AP, Hume ID (eds) Possums and gliders. Australian Mammal Society, Sydney, pp 191–196

Gates GR, Saunders JC, Bock GR, Aitkin LM, Elliott MA (1974) Peripheral auditory function in the platypus, *Ornithorhynchus anatinus*. J Acoust Soc Am 56: 152–156

Gemmel RT, Nelson J (1992) Development of the vestibular and auditory system of the Northern native cat. Anat Rec 234: 136–143

Goodman CS, Shatz CJ (1993) Developmental mechanisms that generate precise patterns of neuronal connectivity. Cell 72/Neuron 10 (Suppl): 77–98

Gossampson MA, Kirss A (1991) Effect of pentobarbital and ketamine-xylazine anaesthesia on somatosensory, brainstem auditory and peripheral sensory-motor responses in the rat. Lab Anim 25: 360–366

Gottlieb G (1971) The ontogenesis of sensory function in birds and mammals. In: Tobach E, Aronson LR, Shaw E (eds) The biopsychology of development. Academic Press, New York, pp 67–128

Green S, Marler P (1979) The analysis of animal communication. In: Marler P, Vandenbergh JG (eds) Handbook of behavioral neurobiology vol 3, Social behavior and communication. Plenum, New York, pp 73–158

Gummer AW, Mark RF (1994) Patterned neural activity in brain stem auditory areas of a prehearing animal, the tammar wallaby (*Macropus eugenii*). NeuroReport 5: 685–688

Hack MH (1968) The developmental Preyer reflex in the sh-1 mouse. J Aud Res 8: 449–457

Haight JR, Neylon L (1978) The organization of neocortical projections from the ventroposterior thalamic complex in the marsupial brush-tailed possum, *Trichosurus vulpecula*: a horseradish peroxidase study. J Anat 126: 459–485

Haight JR, Neylon L (1979) The organization of neocortical projections from the ventrolateral thalamic nucleus in the brush-tailed possum, *Trichosurus vulpecula*, and the problem of motor and somatic sensory convergence within the mammalian brain. J Anat 129: 673–694

Harrison JM, Feldman ML (1970) Anatomical aspects of the cochlear nucleus and superior olivary complex. In: Neff WD (ed) Contributions to sensory physiology vol 4. Academic Press, New York, pp 95–142

Heath CJ, Jones EG (1971) Interhemispheric pathways in the absence of the corpus callosum. J Anat 109: 253–270

Heffner HE (1983) Hearing in large and small dogs: absolute thresholds and size of the tympanic membrane. Behav Neurosci 97: 310–318

Heffner H, Masterton B (1975) Contribution of auditory cortex to sound localization in the monkey (*Macaca mulatta*). J Neurophysiol 38: 1340–1358

Heffner HE, Masterton B (1980) Hearing in glires: domestic rabbit, cotton rat, house mouse and kangaroo rat. J Acoust Soc Am 68: 1584–1599

Heffner R, Heffner H (1980) Hearing in the elephant (*Elephas maximus*). Science 208: 518–520

Heffner RS, Heffner HE (1985a) Hearing range of the domestic cat. Hear Res 19: 85–88

Heffner RS, Heffner HE (1985b) Hearing in mammals: the least weasel. J Mammal 66: 745–755

Heffner RS, Heffner HE (1987) Localization of noise, use of binaural cues, and a description of the superior olivary complex in the smallest carnivore, the least weasel (*Mustela nivalis*). Behav Neurosci 101: 701–708

Heffner RS, Heffner HE (1992) Visual factors in sound localization in mammals. J Comp Neurol 317: 219–232

Heffner RS, Masterton RB (1990) Sound localization in mammals: brain-stem mechanisms. In: Berkley MA, Stebbins WC (eds) Comparative perception vol 1. Basic mechanisms. Wiley, New York, pp 285–314

Henkel CK (1981) Afferent sources of a lateral midbrain tegmental zone associated with the pinnae in the cat as mapped by retrograde transport of horseradish peroxidase. J Comp Neurol 203: 213–226

Henkel CK, Edwards SB (1978) The superior colliculus control of pinna movements in the cat: possible anatomical connections. J Comp Neurol 182: 763–776

Hill JP, Hill WCO (1955) The growth stages of pouch young of the native cat (*Dasyurus viverrinus*) together with observations on the anatomy of the newborn young. Trans Zool Soc Lond 28: 349–453

Hood LJ, Berlin CI, Heffner RS, Morehouse CR, Smith EG, Barlow EK (1991) Objective auditory threshold estimation using sine-wave derived responses. Hear Res 55: 109–116

Hopson JA (1966) The origin of the mammalian middle ear. Am Zool 6: 437–450

Hubel DH, Wiesel TN (1970) The period of susceptibility to the physiological effects of unilateral eye closure in kittens. J Physiol (Lond) 206: 419–436

Huff JN, Price EO (1968) Vocalizations of the least weasel, *Mustela nivalis*. J Mammal 49: 548–550

Huxley TH (1880) On the application of the laws of evolution to the arrangement of the Vertebrata, and more particularly of the Mammalia. Proc Zool Soc Lond, pp 649–662

Imig TJ, Brugge JF (1978) Sources and terminations of callosal axons related to binaural and frequency maps in primary auditory cortex of the cat. J Comp Neurol 180: 637–660

Imig TJ, Morel A (1985) Tonotopic organization of ventral nucleus of medial geniculate body in the cat. J Neurophysiol 53: 309–340

Imig TJ, Ruggero MA, Kitzes LM, Javel E, Brugge JF (1977) Organization of auditory cortex of the owl monkey (*Aotus trivirgatus*). J Comp Neurol 171: 111–128

Irvine DRF (1986) The auditory brainstem. In: Ottoson D (ed) Progress in sensory physiology, vol 7. Springer, Berlin Heidelberg New York

Irvine DRF (1992) Physiology of auditory brainstem. In: Popper AN, Fay RR (eds) The mammalian auditory pathway: neurophysiology. Springer, Berlin Heidelberg New York, pp 153–231

Irving R, Harrison JM (1967) The superior olivary complex and audition: a comparative study. J Comp Neurol 130: 77–86

Jane JA, Yashon D, Diamond IT (1968) An anatomic basis for multimodal thalamic units. Exp Neurol 22: 464–471

Jenkins WM, Masterton RB (1982) Sound localization: effects of unilateral lesions in central auditory system. J Neurophysiol 47: 987–1016

Jenkins WM, Merzenich MM (1984) Role of cat primary auditory cortex for sound localization behavior. J Neurophysiol 52: 819–847

Jewett DL (1970) Volume-conducted potentials in response to auditory stimuli as detected by averaging in the cat. Electroencephalograph Clin Neurophysiol 28: 609–618

Johnson JI (1977) The central nervous system of marsupials. In: Hunsacker D (ed) The biology of marsupials. Academic Press, New York, pp 157–278

Johnson JI, Kirsch JAW, Reep RL, Switzer RC III (1994) Phylogeny through brain traits: more characters for the analysis of mammalian evolution. Brain Behav Evol 43: 319–347

Johnstone JR, Alder VA, Johnstone BM, Robertson D, Yates GK (1979) Cochlear action potential threshold and single unit threshold. J Acoust Soc Am 65: 254–257

Jones EG, Rockel AJ (1973) Observations on complex vesicles, neurofilamentous hyperplasia and increased electron density during terminal degeneration in the inferior colliculus. J Comp Neurol 147: 93–118

Jürgens U (1988) Central control of monkey calls. In: Todt D, Goedeking P, Symmes D (eds) Primate vocal communication. Springer, Berlin Heidelberg New York, pp 162–167

Kavanagh RP, Rohan-Jones WG (1982) Calling behaviour of the yellow-bellied glider, *Petaurus australis* Shaw (Marsupialia: Petauridae). Aust Mammal 5: 95–111

Kelly JB, Masterton B (1977) Auditory sensitivity of the albino rat. J Comp Physiol Psychol 91: 930–936

Kelly JB, Kavanagh GL, Dalton JCH (1986) Hearing in the ferret (*Mustelia putorius*). Hear Res 24: 269–275

Kelly JP, Wong D (1981) Laminar connections of the cat's auditory cortex. Brain Res 212: 1–15

Kermack KA, Mussett F, Rigney HW (1981) The skull of *Morganucodon*. Zool J Linn Soc 71: 1–158

Kirsch JAW (1977) The classification of marsupials. In: Hunsacker D (ed) The biology of marsupials. Academic Press, New York, pp 1–50

Kirsch JAW, Calaby JH (1977) The species of living marsupials: an annotated list. In: Stonehouse B, Gilmore D (eds) The biology of marsupials. Macmillan, London, pp 9–26

König N, Roch G, Marty R (1975) The onset of synaptogenesis in rat temporal cortex. Anat Embryol 148: 73–87

Konishi M (1970) Comparative neurophysiological studies of hearing and vocalizations in songbirds. Z Vgl Physiol 66: 257–272

Krubitzer L (1995) The organization of neocortex in mammals: are species differences really so different? Trends Neurosci 18: 408–417

Kudo M, Niimi K (1980) Ascending projections of the inferior colliculus: an autoradiographic study. J Comp Neurol 191: 545–556

Kudo M, Glendenning KK, Frost SB, Masterton RB (1986) Origin of mammalian thalamocortical projections. I. Telencephalic projections of the medial geniculate body in the opossum (*Didelphis virginiana*). J Comp Neurol 245: 176–197

Kudo M, Aitkin LM, Nelson JE (1989) Auditory forebrain organization of an Australian marsupial, the Northern native cat (*Dasyurus hallucatus*). J Comp Neurol 279: 28–42

Kudo M, Kitao Y, Nakamura Y (1990) Differential organization of crossed and uncrossed projections from the superior olive to the inferior colliculus in the mole, Neurosci Lett 117: 26–30

Ladhams A, Pickles JO (1996) Morphology of the monotreme organ of Corti and macula lagena. J Comp Neurol 366: 335–347

Larsell O, McCrady E, Zimmermann AA (1935) Morphological and functional development of the membranous labyrinth in the opossum. J Comp Neurol 63: 95–118

Lawrence BD, Simmons JA (1982) Echolocation in bats; the external ear and perception of the vertical positon of targets. Science 218: 481–483

Lee AK, Cockburn A (1985) Evolutionary ecology of marsupials. Cambridge University Press, Cambridge

Lende RA (1963) Sensory representation in the cerebral cortex of the opossum (*Didelphis virginiana*). J Comp Neurol 121: 395–403

Lent R, Cavalcante LA, Rocha-Miranda CE (1976) Retinofugal projections in the opossum. Anterograde degeneration and radioautographic study. Brain Res 107: 9–26
Lieberman P (1984) The biology and evolution of language. Harvard, Cambridge
Liu GB, Hill KG, Mark RF (1996) The auditory brainstem response (ABR) and responses from brainstem nuclei during development in *Macropus eugenii*. Proc Aust Neurosci Soc 7: 227
Lopez DE, Merchán MA, Bajo VM, Saldaña E (1993) The cochlear root neurons in the rat, mouse and gerbil. In: Merchan MA, Juiz JM, Godfrey DA, Mugnaini E (eds) The mammalian cochlear nuclei: organization and function. Plenum Press, New York, pp 291–301
Lorente de No R (1933) Anatomy of the eighth nerve: the central projections of the nerve endings of the internal ear. Laryngoscope 63: 1–37
Manley GA (1972) A review of some current concepts of the functional evolution of the ear in terrestrial vertebrates. Evolution 26: 608–621
Marler P (1970) Bird song and speech development: could there be parallels? Am Sci 58: 663–669
Marlow BJ (1960) The evolution and radiation of mammals. Aust Mus Mag 13: 184–190
Martin GF (1968) The pattern of neocortical projections to the mesencephalon of the opossum (*Didelphis virginiana*). Brain Res 11: 593–610
Martin GF (1969) Efferent tectal pathways of the opossum (*Didelphis virginiana*). J Comp Neurol 135: 209–224
Martin GF, Bresnaha JC, Henkel CK, Megirian D (1975) Corticobulbar fibers in the North American opossum (*Didelphis marsupialis virginiana*) with notes on the Tasmanian brush-tailed possum (*Trichosurus vulpecula*) and other marsupials. J Anat 120: 439–484
Martin GF, Beals JK, Culbertson JL, Dom R, Goode G, Humbertson AO (1978) Observations of the development of brainstem-spinal systems in the North American opossum. J Comp Neurol 181: 271–289
Marty R, Thomas J (1963) Réponse électro-corticale à la stimulation du nerf cochléaire chez le chat nouveau-ne. J Physiol (Paris) 55: 165–166
Masterton B, Heffner H, Ravizza R (1969) The evolution of human hearing. J Acoust Soc Am 45: 966–985
Masterton RB, Granger EM, Glendenning KK (1992) Psychoacoustical contribution of each lateral lemniscus. Hear Res 63: 57–70
Masterton RB, Granger EM, Glendenning KK (1994) Role of acoustic striae in hearing: mechanism for enhancement of sound detection in cats. Hear Res 73: 209–222
McCrady EJ (1938) The embryology of the opossum. Am Anat Mem 16: 1–233
McCrady E, Wever EG, Bray CW (1937) The development of hearing in the opossum. J Exp Zool 75: 503–517
McCrady E, Wever EG, Bray CW (1940) A further investigation of the development of hearing in the opossum. J Comp Psychol 30: 17–21
McKay GM (1989) Family Petauridae. In: Walton DW, Richardson BJ (eds) Fauna of Australia, Mammalia. Australian Government Printing Services, Canberra, pp 665–678
McManus JJ (1970) Behavior of captive opossums, *Didelphis marsupialis virginiana*. Am Midl Nat 84: 144–169
Meng J, Fox RC (1995a) Therian petrosals from the Oldman and Milk River formations (Late Cretaceous), Alberta, Canada. J Vertebr Paleontol 15: 122–130
Meng J, Fox RC (1995b) Osseous inner ear structures and hearing in early marsupials and placentals. Zool J Linn Soc 115: 47–71
Merchán MA, Collia F, Lopez DE, Saldana E (1988) Morphology of cochlear root neurons in the rat. J Neurocytol 17: 711–725
Merzenich MM (1970) Morphological specialization of the cochlear nuclear complex in certain mammals. Anat Rec 166: 347
Merzenich MM, Brugge JF (1973) Representation of the cochlear partition in the superior temporal plane of the macaque monkey. Brain Res 50: 275–296
Merzenich MM, Schreiner CE (1992) Mammalian auditory cortex – some comparative observations. In: Webster DB, Fay RR, Popper AN (eds) The evolutionary biology of hearing. Springer, Berlin Heidelberg New York, pp 673–689
Merzenich MM, Kitzes L, Aitkin L (1973) Anatomical and physiological evidence for auditory specialization in the mountain beaver (*Aplodontia rufa*). Brain Res 58: 331–344

Merzenich MM, Knight PL, Roth GL (1975) Representation of cochlea within primary auditory cortex in the cat. J Neurophysiol 38: 231–249

Moore DR (1982) Late onset of hearing in the ferret. Brain Res 253: 309–311

Moore DR (1990) Auditory brainstem of the ferret: early cessation of developmental sensitivity of neurons in the cochlear nucleus to removal of the cochlea. J Comp Neurol 302: 810–823

Moore DR, Irvine DRF (1981) Plasticity of binaural interaction in the cat inferior colliculus. Brain Res 208: 198–202

Moore DR, Semple MN, Addison PD, Aitkin LM (1984) Properties of spatial receptive fields in the central nucleus of the cat inferior colliculus. I Responses to tones of low intensity. Hear Res 13: 159–174

Morel A, Imig TJ (1987) Thalamic projections to fields A, AI, P and VP in the cat auditory cortex. J Comp Neurol 265: 119–144

Morest DK (1964) The neuronal architecture of the medial geniculate body of the cat. J Anat Lond 98: 611–630

Morest DK (1965) The laminar structure of the medial geniculate body of the cat. J Anat Lond 99: 143–160

Morest DK, Oliver DL (1984) The neuronal architecture of the inferior colliculus of the cat: defining the functional anatomy of the auditory midbrain. J Comp Neurol 222: 209–236

Morest DK, Winer JA (1986) The comparative anatomy of neurons: homologous neurons in the medial geniculate body of the opossum and cat. Adv Anat Embryol Cell Biol 97: 1–96

Morton SR, Dickman CR, Fletcher TP (1989) Dasyuridae. In: Walton DW, Richardson BJ (eds) Fauna of Australia. Mammalia. Australian Government Publishing Service, Canberra, pp 560–582

Müller M, Wess F-P, Bruns V (1993) Cochlear place-frequency map in the marsupial *Monodelphis domestica*. Hear Res 67: 198–202

Neff WD (1961) Neural mechanisms of auditory discrimination. In: Rosenblith WA (ed) Sensory communication. Wiley, New York, pp 259–278

Nelson JE (1988) Growth of the brain. In: Tyndale-Biscoe CH, Janssens PA (eds) The developing marsupial. Models for biomedical research. Springer, Berlin Heidelberg New York, pp 86–100

Nelson JE, Stephan H (1982) Encephalization in Australian marsupials. In: Archer M (ed) Carnivorous marsupials. Royal Zoological Society of NSW, Sydney, pp 699–706

Neylon L, Haight JR (1983) Neocortical projections of the suprageniculate and posterior thalamic nuclei in the marsupial brush-tailed possum, *Trichosurus vulpecula*, (Phalangeridae), with a comparative commentary on the organization of the posterior thalamus in marsupial and placental mammals. J Comp Neurol 217: 357–375

Northcutt RG, Kaas JH (1995) The emergence and evolution of mammalian neocortex. Trends Neurosci 18: 373–379

Okado N (1981) Onset of synapse formation in the human spinal cord. J Comp Neurol 201: 211–219

Oliver DL (1985) Quantitative analysis of axonal endings in the central nucleus of the inferior colliculus and distribution of ^{3}H-labeling after injections in the dorsal cochlear nucleus. J Comp Neurol 237: 343–359

Oliver DL, Morest DK (1984) The central nucleus of the inferior colliculus in the cat. J Comp Neurol 222: 237–264

Oswaldo-Cruz E, Rocha-Miranda CE (1968) The brain of the opossum (*Didelphis marsupialis*). Instituo de Biofisica, Universidade do Rio de Janeiro, Rio de Janeiro

Pettigrew J (1979) Binocular visual processing of the owl's telencephalon. Proc R Soc Lond Ser B 204: 435–454

Phillips DP, Irvine DRF (1979) Acoustic input to single neurons in pulvinar-posterior complex of cat thalamus. J Neurophysiol 42: 123–136

Ploog D (1988) Neurobiology and pathology of subhuman vocal communication and human speech. In: Todt D, Goedeking P, Symmes D (eds) Primate vocal communication. Springer, Berlin Heidelberg New York, pp 195–212

Poole JH, Payne K, Langbauer WR, Moss CJ (1988) The social context of some very low frequency calls of African elephants. Behav Ecol Sociobiol 22: 385–392

Populin LC, Yin TCT (1995) The topographical organization of the motoneuron pools that innervate the muscles of the pinna of the cat. J Comp Neurol 363: 600–614

Pratt H, Sohmer H (1978) Comparison of hearing threshold determined by auditory pathway electric responses and by behavioral responses. Audiology 17: 285–292

Pujol R (1985) Morphology, synaptology and electrophysiology of the developing cochlea. Acta Otolaryngol Suppl 421: 5–9

Pujol R, Hilding D (1973) Anatomy and physiology of the onset of auditory function. Acta Otolaryngol 76: 1–10

Pujol R, Lavigne-Rebillard M, Lenoir M (1997) Development of sensory and neural structures in the mammalian cochlea. In: Rubel E, Popper AN, Fay RR (eds) Development of the auditory system. Handbook of auditory research, vol 5. Springer, Berlin Heidelberg New York (in press)

Pye JD (1968) Hearing in bats. In: de Reuck AVS, Knight J (eds) Hearing mechanisms in vertebrates. Churchill, London, pp 66–88

Pysh JJ (1969) The development of the extracellular space in neonatal rat inferior colliculus: an electron microscopic study. Am J Anat 124: 411–430

Pysh JJ (1970) Mitochondrial changes in rat inferior colliculus during postnatal development: an electron microscopic study. Brain Res 18: 325–342

Rajan R (1990) Functions of the efferent pathways to the mammalian cochlea. In: Rowe M, Aitkin L (eds) Information processing in the mammalian auditory and tactile systems. Liss, New York, pp 81–96

Rajan R, Irvine DRF, Cassell JF (1991) Normative N_1 audiogram data for the barbiturate-anaesthetised domestic cat. Hear Res 53: 153–158

Rasmussen GL (1946) The olivary peduncle and other fiber projections of the superior olivary complex. J Comp Neurol 84: 141–220

Ravizza RJ, Masterton B (1971) The habituation of auditory reflexes in the decorticate opossum (*Didelphis virginiana*). Physiol Behav 6: 717–722

Ravizza RJ, Masterton B (1972) Contribution of neocortex to sound localization in opossum (*Didelphis virginiana*). J Neurophysiol 35: 344–356

Ravizza RJ, Heffner HE, Masterton B (1969) Hearing in primitive mammals. I. Opossum (*Didelphis virginiana*). J Aud Res 9: 1–7

Rayleigh Lord (1876) Our perception of the direction of a source of sound. Nature 14: 32–33

Reale RA, Imig TJ (1980) Tonotopic organization in auditory cortex of the cat. J Comp Neurol 192: 265–291

Reale RA, Imig TJ (1983) Auditory cortical field projections to the basal ganglia of the cat. Neuroscience 8: 67–86

Reimer K (1993) Tonotopic organization of the inferior colliculus of the grey short-tailed opossum, *Monodelphis domestica*, as revealed by the 2-DG method. In: Elsner N, Heisenberg M (eds) Gene – brain – behaviour. Thieme, Stuttgart, 266 pp

Reimer K (1995) Hearing in the marsupial *Monodelphis domestica* as determined by auditory evoked brainstem responses. Audiology 34: 334–342

Reimer K (1996) Ontogeny of hearing in the marsupial, *Monodelphis domestica*, as revealed by brainstem auditory evoked potentials. Hear Res 92: 143–150

Reimer K, Baumann S (1995) Behavioral audiogram of the Brazilian grey short tailed opossum, *Monodelphis domestica* (Metatheria, Didelphidae). Zoology 99: 121–127

Reynolds A (1975) The development of binaural responses of units in the inferior colliculus of the neonate cat. BSc (Hons) Thesis, Monash University, Melbourne

Ribak CE, Roberts RC (1986) The ultrastructure of the central nucleus of the inferior colliculus of the Sprague-Dawley rat. J Neurocytol 15: 421–438

Rich TH (1991) Monotremes, placentals and marsupials: their record in Australia and its biases. In: Vickers-Rich P, Monaghan JM, Baird JM, Rich TH (eds) Vertebrate palaeontology of Australasia. Pioneer, Melbourne

Ride WDL (1962) On the evolution of Australian marsupials. In: Leeper GW (ed) The evolution of living organisms. Melbourne University Press, Melbourne, pp 281–306

Ride WDL (1968) On the past, present and future of Australian mammals. Aust J Sci 31: 1–11

RoBards MJ (1979) Somatic neurons in the brainstem and neocortex projecting to the external nucleus of the inferior colliculus. J Comp Neurol 184: 547–566

RoBards MJ, Watkins DW, Masterton RB (1976) An anatomical study of some somesthetic afferents to the intercollicular terminal zone of the midbrain of the opossum. J Comp Neurol 170: 499–524

Robertson D, Irvine DRF (1989) Plasticity of frequency organization in auditory cortex of guinea pigs with partial unilateral deafness. J Comp Neurol 282: 456–471

Rockel AJ, Jones EG (1973a) The neuronal organization of the inferior colliculus of the adult cat. I. The central nucleus. J Comp Neurol 147: 11–60

Rockel AJ, Jones EG (1973b) Observations on the fine structure of the central nucleus of the inferior colliculus of the cat. J Comp Neurol 147: 61–92

Rockel AJ, Heath CJ, Jones EG (1972) Afferent connections to the diencephalon in the marsupial phalanger and the question of sensory convergence in the "posterior group" in the thalamus. J Comp Neurol 145: 105–130

Romand R (1983) Development of the cochlea. In: Romand R (ed) Development of auditory and vestibular systems. Academic Press, New York, pp 47–88

Romand R, Romand MR (1982) Myelination kinetics of spiral ganglion cells in kittens. J Comp Neurol 204: 1–5

Roth GL, Aitkin LM, Andersen RA, Merzenich MM (1978) Some features of the spatial organization of the central nucleus of the inferior colliculus of the cat. J Comp Neurol 182: 661–680

Rowe MJ (1990) Organization of the cerebral cortex in monotremes and marsupials. In: Jones EG, Peters A (eds) Cerebral cortex, vol 8B. Plenum, New York, pp 263–334

Royce GJ, Ward JP, Harting JK (1976) Retinofugal pathways in two marsupials. J Comp Neurol 170: 391–414

Rubel EW (1978) Ontogeny of structure and function in the vertebrate auditory system. In: Jacobsen M (ed) Handbook of sensory physiology, vol IX. Development of sensory systems. Springer, Berlin Heidelberg New York, pp 135–237

Rubel EW, Ryals BM (1983) Development of the place principle: tonotopic organization. Science 219: 514–515

Saldana E, Merchán MA (1992) Intrinsic and commissural connections of the rat inferior colliculus. J Comp Neurol 319: 417–437

Sanderson KJ (1986) Evolution of the lateral geniculate nucleus. In: Pettigrew JD, Sanderson KJ, Levick WR (eds) Visual neuroscience. Cambridge University Press, Cambridge, pp 183–195

Sanderson KJ, Aitkin LM (1990) Neurogenesis in a marsupial: the brush-tailed possum (*Trichosurus vulpecula*). I. Visual and auditory pathway. Brain Behav Evol 35: 325–338

Sanderson KJ, Weller WL (1990) Neurogenesis in a marsupial: the brush-tailed possum (*Trichosurus vulpecula*). II. Sensorimotor pathways. Brain Behav Evol 35: 339–349

Sanderson KJ, Pearson LJ, Dixon PG (1978) Altered retinal projections in brushtailed possum, *Trichosurus vulpecula*, following removal of one eye. J Comp Neurol 180: 841–868

Sanderson KJ, Pearson LJ, Haight JR (1979) Retinal projections in the Tasmanian devil, *Sarcophilus harrisii*. J Comp Neurol 188: 335–346

Sanderson KJ, Haight JR, Pettigrew JD (1984) The dorsal lateral geniculate nucleus of macropodid marsupials: cytoarchitecture and retinal projections. J Comp Neurol 224: 85–106

Sanderson KJ, Nelson JE, Crewther DP, Crewther SG, Hammond VE (1987) Retinogeniculate patterns in diprotodont marsupials. Brain Behav Evol 30: 22–42

Saunders JC, Gates GR, Coles RB (1974) Brain-stem evoked responses as an index of hearing thresholds in one-day-old chicks and ducklings. J Comp Physiol Psychol 86: 426–431

Saunders JC, Doan DE, Cohen YE (1993) The contribution of middle-ear sound conduction to auditory development. Comp Biochem Physiol 106A: 7–13

Saunders NR, Adam E, Reader M, Mollgard K (1989) *Monodelphis domestica* (grey short-tailed opossum): an accessible model for studies of early neocortical development. Anat Embryol 180: 227–236

Schiller PH (1984) The superior colliculus and visual function. In: Darian-Smith I (ed) Handbook of physiology, sec 1: the nervous system, vol III. Sensory processes, part 1. American Physiological society, Bethesda, pp 457–505

Schroeder DM, Jane JA (1976) The intercollicular area of the inferior colliculus. Brain Behav Evol 13: 125–141

Segall W (1969) The auditory ossicles (malleus, incus) and their relationship to the tympanic: in marsupials. Acta Anat 73: 176–191

Segall W (1971) The auditory region (ossicles, sinuses) in gliding mammals and selected representatives of non-gliding genera. Fieldiana Zool 58: 27–59

Seiden HR (1957) Auditory acuity of the marmoset monkey (*Hapale jacchus*). PhD Thesis, Princeton University, Princeton

Seyfarth R, Cheney D (1986) Vocal development in vervet monkeys. Anim Behav 34: 1640–1665

Shatz C (1990) Impulse activity and the patterning of connections during CNS development. Neuron 5: 745–756

Sheng XM, Marotte LR, Mark RF (1990) Development of connections to and from the visual cortex in the wallaby (*Macropus eugenii*). J Comp Neurol 300: 196–210

Short RV (1985) Hopping mad. In: Proc IUPS Congress, Sydney. Australian Academy of Science, Canberra, pp 371–386

Simpson GG (1945) The principles of classification and a classification of mammals. Bull Am Mus Nat Hist 85: 1–350

Simpson GG (1965) Geography of evolution. Chilton, Philadelphia

Smith CA, Takasaka T (1971) Auditory receptor organs of reptiles, birds and mammals. Contrib Sens Physiol 5: 129–177

Smith LE, Simmons FB (1982) Accuracy of auditory brainstem evoked response with hearing level unknown. Ann Otol Rhinol Laryngol 91: 266–267

Smith M (1980) Behaviour of the koala, *Phascolarctos cinereus* (Goldfuss), in captivity. III. Vocalizations. Aust Wildl Res 7: 13–34

Spoendlin H (1984) Primary neurons and synapses. In: Friedmann I, Ballantyne J (eds) Ultrastructural atlas of the inner ear. Butterworths, London, pp 133–164

Sprague JM, Meikle TH (1965) The role of the superior colliculus in visually guided behavior. Exp Neurol 11: 115–146

Stein BE, Clamann HP (1981) Control of pinna movements and sensorimotor register in cat superior colliculus. Brain Behav Evol 19: 180–192

Stephan H (1972) Evolution of primate brains: a comparative anatomical investigation. In: Tuttle R (ed) The functional and evolutionary biology of primates. Atherton, Chicago, pp 155–174

Stokes JH (1912) The acoustic complex and its relations in the brain of the opossum (*Didelphys virginiana*). Am J Anat 12: 401–445

Stonehouse B, Gilmore D (1977) The biology of marsupials. University Park Press, Baltimore

Struhsaker TT (1967) Auditory communication among vervet monkeys (*Cercopithecus aethiops*). In: Altmann SA (ed) Social communication among primates. University of Chicago Press, Chicago, pp 281–324

Suga N (1984) The extent to which biosonar information is represented in the bat auditory cortex. In: Edelman GM, Gall WE, Cowan WM (eds) Dynamic aspects of neocortical function. Wiley, New York, pp 315–373

Syka J, Radil-Weiss T (1971) Electrical stimulation of the tectum in freely moving cats. Brain Res 28: 567–572

Syka J, Straschill M (1970) Activation of superior colliculus neurons and motor responses after electrical stimulation of the inferior colliculus. Exp Neurol 28: 384–392

Szalay FS (1982) A new appraisal of marsupial phylogeny and classification. In: Archer M (ed) Carnivorous marsupials. Royal Zoological Society of New South Wales, Sydney, pp 621–640

Taber-Pierce E (1973) Time of origin of neurons in the brain stem of the mouse. Prog Brain Res 40: 53–65

Todt D, Goedeking P, Symmes D (1988) Primate vocal communication. Springer, Berlin Heidelberg New York

Tonndorf J, Khanna SM (1967) Some properties of sound transmission in the middle and outer ears of cats. J Acoust Soc Am 41: 513–521

Tumarkin A (1968) Evolution of the auditory conducting apparatus in terrestrial vertebrates. In: de Reuck AVS, Knight J (eds) Hearing mechanisms in vertebrates. Churchill, London, pp 18–40

Tyndale-Biscoe H (1973) Life of marsupials. Arnold, London

Ulinski P (1983) Dorsal ventricular ridge. Wiley, New York

von Békésy G (1960) Experiments in hearing. McGraw-Hill, New York

Wada T (1923) Anatomical and physiological studies of the growth of the inner ear of the albino rat. Wistar Inst Anat Biol 10: 1–74

Walker EP (1968) Mammals of the world. Johns Hopkins Press, Baltimore

Warr WB (1992) Organization of olivocochlear efferent systems in mammals. In: Webster DB, Popper AN, Fay RR (eds) The mammalian auditory pathway: neuronanatomy. Springer, Berlin Heidelberg New York, pp 41–448

Wever EG (1974) The evolution of vertebrate hearing. In: Keidel WD, Neff WD (eds) Handbook of sensory physiology, vol 1. Auditory system. Anatomy physiology (ear). Springer, Berlin Heidelberg New York, pp 423–454

Wiener FM, Pfeiffer RR, Backus ASN (1966) The sound pressure transformation by the head and auditory meatus of the cat. Acta Otolaryngol 61: 255–269

Willard FH (1993) Postnatal development of auditory nerve projections to the cochlear nucleus in *Monodelphis domestica*. In: Merchán MA, Juiz JM, Godfrey DA, Mugnaini E (eds) The mammalian cochlear nuclei: organization and function. Plenum Press, New York, pp 29–42

Willard FH, Martin GF (1983) The auditory brainstem nuclei and some of their projections to the inferior colliculus in the North American opossum. Neuroscience 10: 1203–1232

Willard FH, Martin GF (1984) Collateral innervation of the inferior colliculus in the North American opossum: a study using fluorescent markers in a double-labeling paradigm. Brain Res 303: 171–182

Willard FH, Martin GF (1986) The development and migration of large multipolar neurons into the cochlear nucleus of the North American opossum. J Comp Neurol 248: 119–132

Willard FH, Munger BL (1988) Sequential waves of differentiation in the cochlea of *Monodelphis domestica*. Soc Neurosci 14: 425

Willott JF, Aitkin LM, McFadden SL (1993) Plasticity of auditory cortex associated with sensorineural hearing loss in adult C57BL/6J mice. J Comp Neurol 329: 402–411

Winer JA (1985) The medial geniculate body of the cat. Adv Anat Embryol Cell Biol 86: 1–98

Winer JA (1992) The functional architecture of the medial geniculate body and the primary auditory cortex. In: Webster DB, Popper AN, Fay RR (eds) The mammalian auditory pathway: neuroanatomy. Springer, Berlin Heidelberg New York, pp 222–409

Winter JW (1976) The behaviour and social organization of the brushtailed possums (*Trichosurus vulpecula*: Kerr). PhD Thesis, University of Queensland, Brisbane

Withington DJ, Mark RF, Thornton SK, Liu GB, Hill KG (1995) Neural responses to free-field auditory stimulation in the superior colliculus of the wallaby (*Macropus eugenii*). Exp Brain Res 105: 233–240

Woodburne MO (1984) Families of marsupials: relationships, evolution and biogeography. In: Broadhead TM (ed) Mammals: notes for a short course. Univ Tenn Dept Geol Sci Stud Geol 8y, Nashville

Woolsey CN (1960) Organization of cortical auditory systems: a review and a synthesis. In: Rasmussen GL (ed) Neural mechanisms of the auditory and vestibular systems. Thomas, Springfield, pp 165–180

Xu SA, Shepherd RK, Chen Y, Clark GM (1993) Profound hearing loss in the cat following the single co-administration of kanamycin and ethacrynic acid. Hear Res 70: 205–213

Subject Index

Springer
and the
environment

At Springer we firmly believe that an international science publisher has a special obligation to the environment, and our corporate policies consistently reflect this conviction.

We also expect our business partners – paper mills, printers, packaging manufacturers, etc. – to commit themselves to using materials and production processes that do not harm the environment. The paper in this book is made from low- or no-chlorine pulp and is acid free, in conformance with international standards for paper permanency.

Springer

If you have any concerns about our products,
you can contact us on
ProductSafety@springernature.com

In case Publisher is established outside the EU,
the EU authorized representative is:
Springer Nature Customer Service Center GmbH
Europaplatz 3, 69115 Heidelberg, Germany

Printed by Libri Plureos GmbH
in Hamburg, Germany